GW00728842

Extrusion of Plastics

Extrusion of Plastics

E.G. FISHER Ph.C., F.P.I.

Assisted by
E. C. WHITFIELD

Published for The Plastics and Rubber
Institute

NEWNES-BUTTERWORTHS
LONDON-BOSTON
SYDNEY-WELLINGTON-DURBAN-TORONTO

THE BUTTERWORTH GROUP

ENGLAND
Butterworth & Co (Publishers) Ltd
London: 88 Kingsway, WC2B 6AB

AUSTRALIA
Butterworths Pty Ltd
Sydney: 586 Pacific Highway, NSW 2067
Also at Melbourne, Brisbaine, Adelaide and Perth

CANADA
Butterworth & Co (Canada) Ltd
Toronto: 2265 Midland Avenue, Ontario MIP 451

NEW ZEALAND
Butterworths of New Zealand Ltd
Wellington: 26-28 Waring Taylor Street, 1

SOUTH AFRICA
Butterworth & Co (South Africa) (Pty) Ltd
Durban: 152-154 Gale Street

USA
Butterworth (Publishers) Inc.
Boston: 161 Ash Street, Reading, Mass. 01867

First published 1958
Second edition 1964
Third edition published in 1976 by Newnes-Butterworths.
Published for The Plastics and Rubber Institute
11 Hobart Place, London SW1W 0HL

ISBN 0 408 00194 1

Typeset by Inforum Ltd, Portsmouth.

Printed in England by Hazell Watson & Viney Ltd, Aylesbury, Bucks.

Preface to Third Edition

The period which elapsed between the publication of the first and second editions of this book was a time of intense activity in the empirical development of new extrusion processes, in the production of new resins for extrusion, and in the greater appreciation and understanding of the extrusion process generally. The years from that time until the date of writing of this third edition seem to have been a period of consolidation in which the principles and processes previously established were developed further and used to the full.

During this latter period machine sizes and outputs increased greatly, as did the number of extrusion lines in operation. Extrusion was no longer a somewhat mysterious novelty but had become a standard tool of plastics fabrication which within the confines of normal plant operation could always be expected to give a product predictable as to output and quality. The 'Art' factor which had featured so greatly in previous years had become of less importance and extrusion had in fact almost become a 'science'.

During this period the use of computers in the design of screws and of dies was becoming possible and the automatic adjustment of production parameters by the same means also became feasible. The rapid development which had occurred in electronics and solid state devices during this period also had its impact on the design of extrusion control systems, to the general improvement of performance and reliability.

Alongside these very important scientific developments the inventors and innovators—always a factor of outstanding importance in the extrusion industry—also continued their work and produced an abundance of new ideas and interesting devices, some of which are already commercially significant whilst others have been abandoned. An exhaustive treatment of all of these innovations would be impossible in one volume, however large, and no attempt has been made to do so in this work.

In this third edition of *Extrusion of Plastics*, therefore, the author has decided not to rewrite any chapter or to add new chapters, since the fundamentals as originally described remain the same, but has selected and added such new developments and innovations as he considers to be worthy of inclusion.

The author wishes to thank his many helpers in the updating of this volume. In particular he wishes to thank the many manufacturers of extrusion equipment who replied to his requests for information on their latest developments and who kindly supplied appropriate photographs for publication.

E.G.F.

Preface to First Edition

The extrusion process, as a means of forming materials, has been in use since the beginning of the 19th century. Although it became an established tool of the rubber and cable industry during the middle or latter part of that century, it was not until the development of the new synthetic polymers that its full potential began to be realised. The rapid growth of the plastics industry and production of the new polymeric materials has given much impetus to the development and perfecting of processes for their manufacture. Extrusion, by virtue of its continuous nature, is one of the most important of such processes.

The book attempts to cover briefly all phases of the extrusion of plastics, from both the practical and theoretical viewpoints, and to treat the more important features in greater detail. The work is intended for the student, but no attempt has been made to deal exhaustively with any aspect of the extrusion process.

In compiling this book, the author has consulted many published articles: in particular those by W. L. Gore and his co-workers of E. I. Du Pont de Nemours & Co. He wishes to thank the authors of these and other publications. He also wishes to record his indebtedness to Tenaplas Limited for permission to refer to their work on pipe extrusion, and to British Resin Products Limited and Mr. D. N. Davies of that Company for assistance with Chapter 8 and permission to publish Fig. 42.

E.G.F.

Preface to Second Edition

The first edition of this book appeared at a time when many developments in the field of extrusion technology and of materials for extrusion were just becoming established. Moreover, during the comparatively short time since the volume was published, there have been a number of significant advances in the plastics industry as a whole and increased use of screw extrusion processes for purposes other than normal continuous extrusion.

The preparation of the second edition therefore has involved many changes. Wherever necessary the existing text has been completely rewritten, several new chapters have been added and in many cases the subject has been treated in greater detail. In particular a chapter has been devoted to thermoplastic processing techniques which are not normally regarded as extrusion operations, but of which screw extrusion forms an essential part. Another new chapter deals with unorthodox extrusion equipment or processes, and 'dies for extrusion' has also been separately dealt with, thus allowing this important subject to be covered in greater detail.

As with the first edition, it has been the author's aim to present a wide field of extrusion technology in an understandable form which he sincerely hopes will be of value to all persons engaged in the industry.

Once again the author wishes to express his thanks to his many helpers without whose assistance this second edition would never have been completed. Particularly he is indebted to Mr E. D. Chard for checking the references and for reading the manuscript, to Mr Colin Whitfield for his valuable criticism and assistance, and for finding sources of information, to Mr A. Kennaway for checking Chapter 3, and his secretaries, in particular Anne Weekes, for typing the manuscript from a collection of scrappy notes. His thanks are due also to the many equipment manufacturers and other firms who co-operated so readily with details and photographs of their latest equipment.

E.G.F.

Contents

1

Introduction

1.1 Historical and general

Although extrusion as a manufacturing process for structural materials can be considered to have originated at the end of the eighteenth century, it is only within the last 30 years that its full potential has begun to be realised. Beginning with Joseph Bramah, who appears to have invented the first extrusion machine[1], the use of the process during the first half of the nineteenth century seems to have been confined to the production of lead pipes. The use of extrusion in one form or another for the manufacture of spaghetti, macaroni, and other foodstuffs, and in the brick-making and ceramics industries, is probably a very old art but the history of these branches of the technique is very obscure.

The first definite stage in the development of extrusion as a manufacturing process occurred in 1845 when Richard Brooman filed a patent for the manufacture and application of gutta percha thread by an extrusion process. This extruder was subsequently modified by Henry Bewley of The Gutta Percha Company, where it was used to cover the copper conductors in the first submarine cable, which was laid between Dover and Calais in 1851[2,3].

In the U.S.A. the first use of extrusion for the manufacture of insulated wires appears to be due to A.G. De Wolfe who, when employed by the A.G. Day Co. of Seymour, Connecticut, commenced work on an extruder for this purpose in 1858, or thereabouts[4].

During the following 25 years the process grew in importance at a rapid rate and mechanically operated presses soon replaced the hand types which had been used previously. Many thousands of kilometres of insulated wires and cables were produced and the extrusion process for cable making was thus firmly established.

The extrusion machines used in those early times, and from which the extrusion industry as it is today has grown, were all of the ram type and were operated either manually, mechanically or hydraulically. In this process hot gutta-percha was pressed by the ram into a die box through which the copper conductor was guided. The gutta-percha was extruded from the orifice of the die box and was thus made to coat the conductor with an insulating layer. Such machines suffered from an obvious disadvantage in that their operation was not continuous and had to stop at regular intervals to refill, or change the cylinder. A number of attempts were made, therefore, to overcome this limitation and these resulted finally in the adoption of the screw principle of extrusion.

It must not be assumed from this that ram extrusion was abandoned. Such machines were still widely used—and are so to this day—for processes where exceptionally high pressures were required or when the materials being extruded were unsuitable for handling by screws. Metals are extruded on ram machines, for example, as are ceramics, graphite, waxes, refractory flux coatings for welding electrodes, and many other products. It must also be remembered that a number of present-day injection moulding machines still use ram extruders as their injection members. The first patent on an extrusion machine employing an Archimedean screw was taken out by Gray in 1879[5], and Royle in the United States also developed a screw machine at approximately the same time (*Figure 1.1*). Shaw in this country had produced, and sold, screw machines by 1881 (*Figure 1.2*), and Iddon, some three years later, designed a double-roller system which was followed in another two years by a screw machine with a right-angled head by the same maker. In Germany the equally well known maker of rubber and plastics machinery, Paul Troester, had produced a successful screw machine by 1892 and by 1912 is believed to have constructed and delivered over 500 such machines (*Figure 1.3*). This was followed shortly afterwards by the publication of an extruder screw design by Phoenix Gummiwerke, Germany[6].

In connection with these early screw extruders, it is interesting to note that the machine developed by A.G. De Wolfe, and referred to previously, was credited in 1872 as having a production speed 1.6 km/h (1 mile/h) of covered cable. Although the size of the cable referred to is not known, it would seem from this that the system used by the inventor was rapid and continuous, and therefore possibly screw-operated. If so—unfortunately there are no further records available—then the credit for producing the first screw machine must go to A.G. De Wolfe. Further details of the history of the extrusion process have been published[6].

Figure 1.1. An early extruder by John Royle dated 1880. (Courtesy John Royle & Sons, Paterson, New Jersey)

4

Figure 1.2. *An early extruder by Francis Shaw & Co. Ltd, dated about 1881. (Courtesy Francis Shaw & Co. Ltd, Manchester)*

Figure 1.3. *An early extruder of German manufacture dated about 1892. (Courtesy Paul Troester Maschinenfabrik, Hannover-Wulfel)*

Apart from gutta-percha the first true thermoplastic—in the now generally accepted meaning of the word—to be processed by extrusion was cellulose nitrate. This material appears to have been extruded for the first time between the years 1875 and 1880 using the wet or solvent-softened cold process in a ram press in much the same operation as is used to this day. Cellulose nitrate was the only extrudable plastic material available in commercial quantities until the early days of the First World War, by which time casein, a protein plastic derived from milk, had assumed industrial importance, and was extruded into rods and tubes for subsequent machining. The British Xylonite Company and Erinoid Ltd were among the earliest firms to undertake this work in the U.K.

Thus from 1845, or thereabouts, until the development of the new synthetic polymers and the foundation of the present plastics industry, the extrusion process was confined to the handling of gutta-percha, rubber, cellulose nitrate, and casein. During the years 1920 to 1930, however, many new thermoplastic resins were being developed and examined from the extrusion point of view. The 10 years commencing about 1925 saw, therefore, the beginning of a number of important changes in the extrusion industry. The small-scale extrusion of cellulose acetate by the cold solvent-softened process—as used for cellulose nitrate—was attracting the attention of firms such as the British Xylonite Company and Erinoid Ltd in this country, whilst the manufacture of flexible rubber-like products from plasticiser-softened polyvinyl chloride (PVC) resins—Mipolam—was being investigated in Germany[7,8]. During the same period, and also in Germany, the new resin polystyrene was being experimented with because of its outstanding electrical properties, and rigid pipes from unsoftened vinyl resins were being manufactured on an experimental scale. It is interesting to note that, whilst rigid articles from unplasticised pure PVC resin were not produced until 1935, satisfactory products were, in fact, manufactured from unplasticised 'mixed' polymer products of vinyl chloride and acrylic ester—Troluloid, later Astralon—as early as 1930[9]. During the latter part of this ten-year period the hot dry extrusion of cellulose acetate became a commercial possibility.

Up to this time the extrusion machines used for both experimental work and production were either ram-operated or were of a screw type, differing little from the early machines which were designed for gutta-percha and rubber and patented by Gray in 1879. The heating medium on these machines was invariably steam, which was soon found to have limitations when used with the new polymers, and the screws were too short to allow adequate heating. The feeding arrangements were designed for hot strip stock which was also inconvenient with the new materials.

As with the early screw extruders for rubber, the history of the development of machines for plastics extrusion is also somewhat obscure. The cloak of mystery which surrounds these early events is the result of the secrecy which was so much a part of the early extrusion industry. Operators were sworn to secrecy and visitors were not allowed. It is known also that many manufacturers for competitive reasons designed and built their own equipment, which was seldom patented.

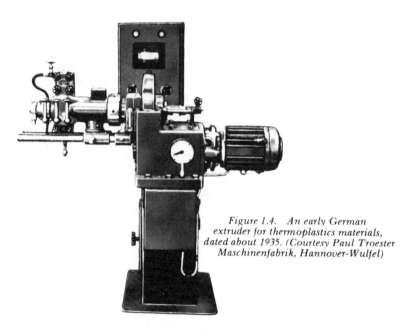

Figure 1.4.   An early German extruder for thermoplastics materials, dated about 1935. (Courtesy Paul Troester Maschinenfabrik, Hannover-Wulfel)

One of the earliest commercially available screw extruders, which had been specially designed for thermoplastics materials however, appears also to have been due to Paul Troester, who offered such a machine from 1935 or thereabouts (*Figure 1.4*). At about the same time Horst Heidrich of Berlin produced his first extruder for plastics materials (*Figure 1.5*) and a few years later, 1937-38, Francis Shaw in this country also produced such a machine (*Figure 1.6*). These new machines were equipped with longer screws and were heated either by oil as a transfer medium, or directly by electrical resistance heaters. In some cases they were provided with variable-speed drives and were arranged for either cold or preheated chip feed. In 1939 Paul Troester took the development a stage further by offering a machine equipped with direct electrical heating, air cooling, automatic temperature control using separate control cabinet, variable-speed drive using a PIV gear, nitrogen-hardened barrel liner, and a screw with a 10 : 1 L/D ratio. This machine (*Figure 1.7*), can thus be looked upon as the true forerunner of present day single-screw extruders, an example of which is given for comparison in *Figure 1.8*.

Figure 1.5.   An early extruder for thermoplastics by Horst Heidrich, Berlin, dated about 1936

   The 1930s are also of particular interest because it is believed to be during this period that the twin-screw principle was being adapted for the extrusion of thermoplastics by Roberto Colombo and Carlo Pasquetti in Italy, to become the forerunners of this now important branch of the extrusion industry.

Figure 1.6.   An early English extruder for thermoplastics. (Courtesy Francis Shaw & Co. Ltd, Manchester)

Figure 1.7.   Highly developed extruder for thermoplastics, dated about 1939. (Courtesy Paul Troester Maschinenfabrik, Hannover-Wulfel)

From these early beginnings the development of the present-day machines has taken place in logical and more or less orderly stages. The improvement of existing polymers and the development of new ones have been followed at each step by an increased interest in their extrusion possibilities. This has called for a continually increasing degree of engineering perfection in the construction of the machines and a greater knowledge of their principles of working. Similarly, the growth of competition in the production and sale of extruded products has caused the designers of extrusion machines to aim at attaining increased outputs from their equipment whilst the extrusion firms have been compelled to produce a wider range of high quality products at low cost.

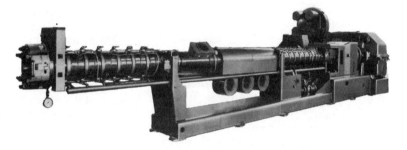

Figure 1.8. Modern extruder for thermoplastics; 200 mm (8 in) diameter, 33 : 1 L/D ratio vented up to 750 kW (1000 hp) thyristor drive, with covers removed showing heaters, etc. (Courtesy Kaufman S.A., Gaillon, France)

1.2 Growth and development of the thermo-plastics extrusion industry

So far as Great Britain is concerned the thermoplastics extrusion industry can be considered in a general way as being divided into two large groups, viz. the cable makers and the specialist extrusion firms. Although the former group—the cable makers—had been operating extrusion machines since 1850 or earlier it was not until comparatively late years that they adopted the newer thermoplastics materials to any substantial degree. The specialist extrusion firms on the other hand naturally founded their branch of the industry on the new materials and have grown with the increasing availability and types of new extrudable resins, and with the development of new applications for them.

1.2.1 THE CABLE MAKERS

The considerable differences between the extrusion conditions which are suitable for rubber as against those required for the new thermoplastics undoubtedly accounted, in part, for the divergence between these two groups, as did the high price of plasticised vinyl materials compared with that of natural rubber. The extrusion plant which was in existence in the cable-making industry was equipped for use with rubber and the operators were familiar with this material. It is not surprising, therefore, that the cable makers did not take a very active part in the early development of plasticised vinyl resins. Likewise, the rubber industry generally did not become actively interested in the new materials until they were more generally accepted. It was not until 1939 and the early war years, with the consequent shortage of natural rubber, that cable makers finally adopted plasticised PVC as a cable insulant. The development of polyethylene with its outstanding electrical properties and excellent extrudability also acted as an effective stimulant in this direction. Any early prejudices which may have existed against the new thermoplastics were overcome as raw materials improved. The cable makers adapted part of their plant to suit the new conditions and soon realised that thermoplastics materials had advantages in many ways over natural rubber. It was found, for example, that whereas with rubber the extrusion machine needed to be set up adjacent to a mill for hot strip feed, plasticised PVC could be fed cold to the machines in the form of a previously compounded chip so that the extrusion and compounding departments could be separate and more compact and thus more economically run. They also found the extrusion of plasticised PVC to be a cleaner operation than that of rubber and, in addition, there was the important advantage of unlimited colour possibilities.

These and other important factors, such as greater resistance to weathering, etc., have had such an effect on the cable industry that to-day the manufacture of insulated wires and cables is in volume one of the most important uses for plasticised PVC and polyethylene.

1.2.2 THE SPECIALIST EXTRUSION FIRMS

Although this branch of the plastics extrusion industry is today concerned with a very wide range of extruded products, the majority of which have no connection with the cable industry, it is interesting to note that it did in fact start with the production of insulated wires and sleevings. In Great Britain the first of the new thermoplastics materials to be commercially available was the plasticised PVC imported from Germany under the trade name Mipolam and the most obvious use for this material was that of flexible electrical insulation. It was natural, therefore, that the first extrusion machines to be set up in this country should be used for this purpose.

On 1 April 1933, Mr B.H. Hornung, the founder and present Managing Director of Suflex Ltd, settled in this country from Berlin, where he had had a successful business in the manufacture of varnished textile insulating sleeving, and had already commenced the production of extruded PVC sleeving and insulated wires. Mr Hornung commenced his business on similar lines in this country, and by 1935 was again producing extruded PVC sleeving and covered wires, using imported Heidrich extruders, in addition to varnished textile sleeving. In 1936 Mr Hornung engaged the services, also from Germany, of a Mr Julius Veit, a name which was to become equally well known in the history of the British extrusion industry[10].

Apart from the British Xylonite Company and Erinoid Ltd, who were by this time producing cellulose acetate tube on their casein extruders, the next name which comes to light is that of Hydroplastics Ltd. This firm was formed during 1935 or 1936, ostensibly for the manufacture of a new type of reinforced phenolic laminate. Hydroplastics Ltd was, however, also interested in other new processes generally and at a later date engaged Veit—who, in the meantime, had left Suflex Ltd—to develop the extrusion of thermoplastics. With the approach of the war the interest of Hydroplastics Ltd in laminated materials decreased whilst their development of extruded sleeving, etc., went on apace. Finally, Hydroplastics Ltd decided to give their entire attention to extrusion and to abandon their original projects. To cover this changed outlook a new company was formed around the old organisation and was called Tenaplas Ltd, a name which has since become well known in the field of thermoplastics extrusion.

The final step in the history of the plastics extrusion industry in Great Britain was taken when Veit decided to start an extrusion firm on his own account and formed Duratube & Wire Ltd during the early war years.

It is also interesting to note, and perhaps a sign of the times, that these two firms Tenaplas Ltd and Duratube & Wire Ltd, once fierce competitors, have now joined forces and are under the same control.

Since those early days there have been many important newcomers to the field of thermoplastics extrusion and an even greater increase in the number of applications for extruded products. The specialist extrusion firms have themselves gradually become divided into two large groups depending on the final destination and usage of the extruded products they manufacture.

1.2.2.1   *Custom extrusion firms*

The organisations coming within this group manufacture extrusions of any type as required for supply in bulk to other manufacturers or distributors who either incorporate them into their own particular products as component parts or sell the extrusions in small quantities individually packed for retail distribution.

The equipment used by custom extrusion firms and their personnel must, therefore, be very versatile and capable of undertaking any extrusion work in any material at short notice in accordance with market requirements.

1.2.2.2   *Special product extrusion firms*

With the growth of the extruded plastics industry generally, certain special products have arisen whose importance has, in time, become sufficiently great to justify the setting up of specialist extrusion plant for their sole manufacture.

The manufacture of insulated wires and cables could be considered in this light as the plant has been specially developed for this specific purpose and is seldom used for the manufacture of other extruded products.

Of recent years polyethylene pipe, film and coated paper represent other examples of such special products, as do rigid PVC pipe and rainwater goods, PVC corrugated roofing and polystyrene and PVC sheet for vacuum forming. These special products, the list of which is growing daily, are usually produced on special extrusion plants which are often purchased complete as 'packaged units' by firms whose extrusion interest lies only in the manufacture of the one product. As would be expected from such specialisation, the equipment used is often elaborate and expensive, and is capable of high outputs with more economic production than would be possible for a custom extrusion firm with its versatile general-purpose plant.

Special-product extrusion firms often handle the marketing of their own products, but they may also supply in bulk to other firms for finishing operations and for marketing in small quantities. Another branch of this group exists in which the extrusion of a special product is carried out by a department of a large organisation which in the main has little or no interest in the plastics industry. Such organisations usually utilise the whole of the output of their extrusion departments as component parts of manufactured products which may or may not be of a substantially non-plastics nature.

Thus it seems that the present trend in the extrusion industry is for the custom extrusion firms to develop specific products in response to customer demand or as a result of their own search for novel products. When, however, these individual products begin to assume substantial importance the customer often takes over the manufacture himself and develops the process into a specialist undertaking. Alternatively the custom extrusion firm may set up its own special operation or dispose of the rights of the development. Such a trend is a valuable one as it results in the development of the most economic production methods, giving lower prices and, therefore, the wider application of extruded thermoplastics.

REFERENCES

1   ANON., 'The History of Extrusion', *Plastics, Lond.*, **18**, 404 (1953)
2   BEWLEY, H., Br. Pat. 10 825 (1845)
3   BROOMAN, R.A., Br. Pat. 10 582 (1845)
4   HOVEY, V.M., Lecture Delivered at the Annual Convention of The Wire Association, 15 November 1960, Chicago, Illinois
5   GRAY, M., Br. Pat. 5056 (10 December 1879)
6   KAUFMAN, M., 'The Birth of the Plastics Extruder', *Plast. Poly.*, **37**, 243 (1969)
7   FIKENTSCHER, M. and HEUCF, C., Germ. Pat. D.R.P. 654 989 (1930)
8   FIKENTSCHER, M. and WOLFF, W., Germ. Pat. D.R.P. 660 793 (1931)
9   FIKENTSCHER, M., private communication to the author
10   HORNUNG, B.H., private communication to the author

2

Basic Principles of Extrusion

2.1 General

Fundamentally, the process of extrusion consists of converting a suitable raw material into a product of specific cross section by forcing the material through an orifice or die under controlled conditions. In order that this simple conception can have practical value, however, there are certain requirements which must be satisfied concerning both the equipment and the raw material. The equipment must be capable of providing sufficient pressure continuously and uniformly on the material, and in some cases must also have means of softening or otherwise conditioning the material so that it becomes extrudable. The material must be such that when suitably conditioned it will flow under pressure and will solidify when these conditions are removed or, alternatively, can be made to solidify as a result of some chemical change which can be continuously brought about.

A simple analogy to illustrate these points is found in the ordinary tube of toothpaste. The equipment—the metal tube operated by the pressure of the fingers—is able to give pressure to the material by reducing the volume of the tube. The material, which has in effect been conditioned by the addition of softening agents, extrudes through the orifice in the tube end—the 'die'—and retains its shape. Doubtless, if the conditioning agents could be continuously removed from the stream of toothpaste as it left the 'die' it would set into a hard dry rod.

In commercial extrusion there are three general types of mechanism in common use, i.e.

Ram and cylinder
Pumps of various types
Rotating screws

and, in a general survey of extrusion for plastics materials, it is usual to consider the technique as being divided into three classes, i.e. 'wet extrusion', 'spinning or spinneret extrusion' and 'dry extrusion'.

This monograph is chiefly concerned with dry extrusion on screw machines, but no general work on extrusion would be complete without referring briefly to the other mechanisms.

2.2 Ram extrusion

As touched upon briefly in Chapter 1, the first extrusion machines used rams or pistons operating in cylinders on batches of prepared raw material. Such machines are still widely used and for a number of purposes possess certain advantages. The pressure, for example, can be as high as required, is always uniform, and can be controlled to a nicety. The raw material is not subjected to any violent agitation or mixing when in the cylinder—a very important matter with some materials—and the equipment is basically very simple. Any raw material for extrusion, therefore, which can be prepared in a plastic state—or in such a state that it can be rendered plastic by pressure alone—and therefore requires no further conditioning in the press, is usually extruded by ram. Similarly materials such as PTFE, which have no proper flow but must be pressure-compacted, may also be processed by this means. Such machines can be constructed to give high outputs at low capital cost and have few maintenance problems.

Some idea of the function and use of a ram machine can be gained from a description of the Barwell batch extruder introduced in 1961 by Barwell Engineering Ltd. This machine has a 250mm (10 in) ram diameter and 1.5 m (60 in) barrel length. The ram is hydraulically operated and has a rapid approach and return action to increase operating speeds. The maximum batch capacity is 45-65 kg (100-150 lb) according to the specific gravity of the compound but being a ram extruder it will operate successfully on a charge of 4.5 kg (10 lb). The maximum extrudate cross-sectional area is 160 cm² (25 in²).

The low power consumption/production rate ratio which is one of the features of ram extruders is well illustrated in that the Barwell machine uses only a 7.5 kW (10 hp) motor as its energy source. Ram pressure can, of course, be controlled.

Some examples of specific uses for ram extrusion processes have been given in Chapter 1 and others will occur from time to time in the body of this work.

As the normal press works on a cake, or dolly, of compacted and substantially static material it is self-evident that no appreciable quantity of heat can be given to a non-metallic or other bad heat-conducting material whilst it is in the press. The normal ram process has a certain disadvantage, therefore, in that it can only work with such materials when they are softened in some manner other than by heat, or with those which can be pre-heated to extrusion temperature. In the latter case the extruder barrel and die are fitted with some method of heating, the function of which is to maintain the temperature of the material rather than to increase it.

Another equally apparent disadvantage of the ram extrusion press is its discontinuous operation. When the ram has reached the end of its power stroke the extrusion stops and the ram must be withdrawn so that a further dolly can be inserted. In the case of materials which are thermally or otherwise unstable the cylinder, die orifice, and other parts, must also be carefully cleaned, removing any scraps of hardened material left behind from the previous stroke. Various ingenious schemes for simplifying and speeding up this operation have been developed[1,2].

Production of rod, strip, sheet, tube, profile in suitable materials, is carried out on ram extruders and it is often the practice to use multi-aperture dies to offset the low production rates of small sections. The recently developed plastics working process known as 'cold forging' is to some extent allied to ram extrusion. Some details of this process have been published[3].

2.3   Wet extrusion

As far as the plastics industry is concerned, the most important remaining use for ram extrusion machines—apart from the older type of injection moulding press which is a special case—is in the so-called 'wet' process, which has already been referred to. In wet extrusion, the raw material is conditioned or softened by the prior addition of solvents, so that little or no further softening—by means of heat for example—is required during extrusion. In a few cases this method has certain advantages over 'dry' extrusion, where softening is obtained by heat and pressure only.

With a material such as cellulose nitrate, which is highly inflammable and dangerous when overheated, it is advisable for all manipulative operations to take place in a wet condition, at relatively low temperatures and with a minimum of compressive effort and frictional effects. From the time they were first extruded, therefore, cellulose nitrate plastics have been processed in the form of a solvent-containing dough or jelly. The frictional effects which are a characteristic of normal screw extrusion could form a dangerous hazard with cellulose nitrate, even if it is in the form of a solvent-softened mixture. Ram presses, therefore, are still used for this material, although machines with very slowly rotating screws acting as low pressure pumps are also used[4].

When cellulose acetate first became available in commercial quantities it was natural that it should be processed in the same manner as the nitrate[5]. Solvent-containing mixtures were, therefore, made up into a dough with suitable plasticisers and extruded directly in this form from a ram press. As with cellulose nitrate and other wet extrusions, the product required to be thoroughly dried and seasoned before use in order that the solvents should be completely removed.

The wet extrusion process, whether carried out on a ram press or otherwise, is still thought by some people to have a certain advantage in the extrusion of cellulose acetate. This thermoplastic is a substantially natural product which often contains impurities and hard particles. The process of acetylation does not always produce a uniform product, therefore, and the dry extrusion process can do little to overcome this. Wet extrusion, because of the softening action and ripening effect of the solvents, can produce a more uniform material which, when extruded, may have a superior finish. Some extruded products in cellulose acetate are still occasionally produced by the wet process, therefore, although with the continual improvement in raw materials their number is now quite small.

Wet extrusion has, on the other hand, certain obvious disadvantages. A special operation is required in the preparation of the raw material and the solvents used are often highly inflammable and expensive. The finished product cannot become dimensionally stable until all traces of solvent have been removed and this often takes a long time, particularly with thick sections. For economic production the installation of an expensive solvent recovery plant is usually essential.

Although, as it has been pointed out, normal wet extrusion is usually carried out on ram presses, there is no fundamental reason why screw machines should not also be used. One of the main features of the screw machine, however—as will be explained—is its ability, by simultaneous mixing and heating under pressure, to bring the material to a suitable condition for extrusion. With wet extrusion this feature would be largely wasted and might in fact, bring complications due to feed difficulties, and to unnecessary agitation of the stock.

Wet extrusion is also carried out with rotary pumps, particularly in the man-made fibre industry, as described in the following section.

2.4   Extrusion pump and spinneret extrusion

Low-viscosity materials, such as solutions and some melts, can be extruded at comparatively low pressures so that substantially standard pumps of one form or another may be used. This system is adopted for the extrusion of monofilaments in regenerated cellulose, cellulose acetate, nylon, and other synthetic fibre-forming materials, where a solution or low-viscosity melt is filtered and pumped to a special multiple die called a spinneret[6,7].

This spinneret contains a large number of very fine holes through which the filtered fibre-forming material is forced by the action of the pump. The extrusion is usually arranged in a downward direction and the monofilaments, which are initially in the form of fine jets of fluid, are caused to solidify in various ways, depending on the requirements of the material. A melt, for example, will set up merely by the action of cooling, whereas solutions require special equipment to drive off—and reclaim—the solvent. Regenerated cellulose and some other monofilaments are extruded directly into a coagulation bath where the required characteristics of the fibres are developed.

The fibres produced in the above manner are subjected to a number of further processes in order to arrive at a final product suitable for weaving and other end-uses. Such techniques are, however, outside the scope of this book although further information is given in sub-section 9.8 'Monofilament Plant', Chapter 9.

Extrusion from solution by means of a pump is also commonly used in other industries: for example, sausage casings in regenerated cellulose and other materials are usually produced by a technique which is roughly similar to that used for viscose fibres. Likewise, the process for the manufacture of regenerated cellulose film has also much in common.

2.5   Dry extrusion

The wet extrusion process described above relies on solvents to soften or plasticise the material so that it becomes extrudable. In the dry process heat alone is used to soften the stock. Dry extrusion may use a ram press in which the material is separately preheated to the required degree before pressing or it may be forced by the ram through a special heating chamber, as in the ram-type injection moulding machine.

The most important method of dry extrusion, however, uses a screw mechanism in which the material enters in a cold form and is then softened by heating and at the same time compacted and brought under pressure by a screw. A combination of screw and ram where the screw has an axial movement is now employed extensively for injection moulding, as is described in greater detail in a later chapter.

2.6 Screw extrusion

2.6.1 BASIC PRINCIPLES

Fundamentally, the screw extrusion machine consists of a screw of special form rotating in a heated barrel or cylinder in which a feed opening is placed radially or tangentially at one end, and an orifice or die axially at the other. A restriction in the form of a breaker plate and screen is sometimes placed between the end of the screw and the extruding die in order to assist the build-up of a pressure gradient along the screw. Recent techniques have indicated that control of pressure at the die is important and a valve is now sometimes used in addition to a breaker plate and screen. The screw is usually bored throughout, or for some part of its length, so that it may be fluid cooled or heated, according to the requirements of the feed material. Screw extrusion can be performed by multiple- as well as single-screw mechanisms. The basic principles remain the same, however, and differences in detail are explained in a later chapter.

The rotating screw takes the material—which is usually in the form of free-flowing cold chips, powders or cubes—from the feed opening, through the heated barrel zones, and compacts it against the breaker plate or other restriction, so that a pressure is built up. During this period the material is forced into intimate and substantially sliding contact with the hot barrel walls and is also sheared and worked so that frictional effects are produced. The combined effects of the hot barrel and the heat due to internal friction in the material cause the thermoplastic to soften so that it may be forced through the restriction to the extrusion die, where it is given the required form.

At this stage it can be assumed that the machine functions by virtue of the friction differential between the screw and the material, as against that between the material and the barrel bore, the latter being the greater[8].

The advantages of a mechanism of this type will be readily apparent as the material is heated uniformly in a closed system and the process is continuous. Considerable mixing also takes place due to the action of the screw and the system is, to some extent, self-compensating in that the heat due to internal friction decreases with the viscosity of the melt.

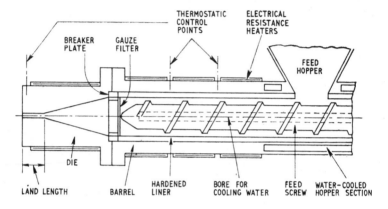

Figure 2.1. Diagram showing essential features of a screw extruder

A simplified sketch of a screw extrusion machine showing the various components is given in *Figure 2.1*. An examination of this will show that there are certain features which can be varied in order to modify the function of the mechanism.

The more important of these features are considered below.

2.6.1.1 Diameter and length of the screw

If the screw is considered solely as a means of conveying the plastics material then the diameter and length are major factors in determining its volumetric capacity and hence the quantity of feed material it can handle. These two functions remain important controls of capacity irrespective of screw design. The length and diameter have a second important influence in that they affect the rate at which heat is transferred from the barrel walls to the material, and this in turn affects the amount of heat generated by friction and shear, the energy input and the power to throughput ratio. It is therefore important that the significance of the length to diameter ratio of screws (*L/D* ratio) which is quoted in specifications be clearly understood. Experimental work in the U.S.A. and Germany has established that long screws, i.e. with *L/D* ratios of 25 : 1 and over, can give higher outputs whilst at the same time maintaining good extrudate quality. There is an indication that even longer screws might give further improved performance if the mechanical difficulties associated with their use were overcome. It is not wise to generalise however, because different thermoplastics materials require different processing conditions, so that maximum screw efficiency for every material with one design of screw is not possible. In this connection it is interesting to note that one manufacturer in the 1960s introduced an extruder with an *L/D* ratio of 3.5 : 1, which successfully extruded a heavy section in rigid PVC[9].

2.6.1.2 The pitch of the screw thread and its flight form

The pitch and helix angle of the flight is, in conjunction with the screw peripheral speed, one of the major factors determining the output of the machine. It also influences to a marked degree the amount of shear applied to the material and the frictional heat generated thereby. The depth of the flight affects also, and to a marked degree, the amount of heat generated by shear, the volumetric. capacity of the screw and the transfer of heat to the material by direct contact with the barrel. Extending the length of the screw allows a longer time for such heat transfer and thus in this respect tends to modify the effect of flight depth.

2.6.1.3 General design

Extruder screws as a general rule employ single-start flights, although two or even more starts are sometimes used. Other screws have been designed with a second flight starting part way along the shaft which by reducing the volumetric capacity of the channel provides a compression on the softened material as it approaches the die. A later Swiss design, which is described in detail in a later chapter, has a second flight of special form and thus has two non-interconnecting channels, one of which functions as a high pressure melt feed whilst the other, whose volumetric capacity progresses to zero, receives the granular material[10].

The alteration of flight depth to obtain a compression on the material as it moves towards the die is standard practice on most screws and is either carried out progressively throughout the length of the screw, or in stages, or by a combination of both methods. The deliberate reduction of the volumetric capacity of the screw channel or channels as outlined above is necessary in order to accommodate the reduction in volume of the material as it becomes fluid and homogeneous and to apply compression to the material so that the channel is completely filled. Alternatively this effect can be obtained by progressively reducing the pitch of the flights without changing their depth or by a combination of both methods. Various other factors such as degree of mixing required, viscosity and transition temperature all influence the design of the screw and as far as possible these factors must be allied to the economic dictates of high output and low power consumption.

2.6.1.4 Speed of the screw

The peripheral speed of the screw is an important variable in the performance of an extruder, not only in the movement of material but also in establishing the amount of heat generated by friction. The output of an extruder does not necessarily increase in direct proportion to the increase in screw speed or power input. An economical speed which gives maximum output per unit of power input therefore is used as the operating speed. This will, of course, vary with different materials and can be adjusted by temperature regulation and screw design.

2.6.1.5 Die, screen and breaker plate

The restriction to the flow of the melt by a breaker plate and screen system, provides two important functions. Firstly, it causes a back pressure to build up in the screw with, as a consequence, a greater mixing movement to the material and also greater shear; secondly, as a screen it ensures that all material passing to the die is in a completely homogeneous state and free from foreign particles. The back-flow and mixing characteristics of a breaker plate and screen are important in particular in the extrusion of dry blends and similar materials of small particle size.

The breaker plate and screen, although obviously reducing output per unit of power, have a beneficial effect on the finished product in that they control the turbulent flow from the screw and 'feed' it to the die in a smoother manner. The precise effect of the breaker plate-screen system will vary with the viscosity of the material.

The effect of die design on output is considerable and is a subject on which generalities can be misleading; recent work suggests that more attention to the flow in a die and its temperature control would greatly increase output without reducing quality.

2.6.1.6 *The temperature of the barrel, screw, die and material*

The viscosity of a thermoplastics melt varies inversely as the temperature. The output of an extrusion machine, therefore, is affected by the temperature of the material. When in the barrel of the machine the thermoplastic is subject to the combined effects of the barrel heat from an external source and the frictional heat due to work. When it is in the die, however, only the external source is available for heating.

Apart from the question of variation of viscosity of the material with temperature and its effects on output, it will also be obvious that all thermoplastics must have an optimum extrusion temperature. If the extrudate is much below this temperature the product will not be homogeneous, and if it is much above then the thermoplastic will either be overheated and may degrade or the extrudate will be too fluid to handle. The method of controlling the temperature is, therefore, a very important feature and involves controlled cooling as well as heating. In fact the temperature of the extrudate and its effect on production rates and product quality has formed the subject of considerable special investigation. One school of thought has suggested that the lowest possible melt temperature should always be used even though the drive power requirements, when operating in this way, would be proportionately higher for the same output. The main advantage of maintaining a low melt temperature is obviously the reduction in thermal degradation of the polymer during processing but other advantages—perhaps equally important in some cases—are the ease of chilling the low temperature extrudate and of preserving its formed shape and dimensions as it emerges from the die[11].

2.6.1.7 *Physical effects and causes*

So far the various aspects of mechanical design and construction and their general effect on output, quality, etc. have been considered briefly. It is relevant and important that the opposite should also be given, i.e. what conditions are influenced by what mechanical or physical feature. In the following list an indication only is given of the more important of such influences, although the majority of design features will directly or indirectly affect most working conditions.

Screw design involves diameter, length, flight depth, flight land width, helix angle, compression ratio and method of obtaining same.

Volumetric capacity is dependent on screw design, principally on diameter, length and flight depth.

Heat transfer efficiency depends on screw design, screw speed and die compression.

Die compression is influenced by die design, breaker plate and screen design, helix angle and flight depth of the screw, and other screw design characteristics.

Frictional heat may be increased or decreased by screw design, particularly the flight depth, and by screw speed.

Viscosity of a thermoplastics material is controlled by the amount of heat transfer and by frictional heat.

Output is dependent on the volumetric capacity of the screw, its speed, the die compression and the viscosity of the material.

Power consumption depends on viscosity, die compression, screw design and screw speed.

From the foregoing it can be understood that the design of an efficient extruder screw is a somewhat complex matter which has therefore been dealt with more fully in a following chapter.

2.6.2   EARLY SINGLE-SCREW EXTRUSION MACHINES

The first screw machines were produced for the processing of rubber, and were intended to receive hot stock in strip form directly from a warming mill. The extrusion machine, therefore, really had little to contribute to the stock in the way of heat and was looked upon merely as a means of masticating and continuously shaping an already heated material by forcing it through a die—in fact, as a continuously acting ram and masticator. The barrel and die were heated chiefly to avoid the lowering of stock temperature by radiation losses.

There has been little significant change in the function and operation of these early machines compared with many of the present day rubber extruders. (Today however it is no longer essential to use preheated stock and the latest rubber extruders will operate from cold strip or granulated stock feed. Such machines have 8 : 1 L/D ratio screws and are sometimes fitted with their own premilling attachments.)

In order to satisfy these requirements, the following features were necessary:

1.   A feed opening, placed tangentially to the screw and equipped with a driven feed roll to facilitate hot strip feed.
2.   A screw and barrel just long enough to compact the material and force it through the breaker plate (if any) and the die.
3.   The barrel to be heated to a relatively low temperature—usually by steam—in order to conserve heat in the material.
4.   A water-cooled feed section and screw, to prevent sticking and pre-curing of the stock respectively.
5.   The feed screw to be deeply cut in order to accept a large volume of material, and designed with decreasing pitch to give a compression ratio without unduly increasing the amount of shear.
6.   The screw to be driven at a relatively low rotational speed in order to avoid excessive frictional heat.
7.   As a result of the above points, the machine need not be particularly robust in its construction and thrust bearing arrangements, and the drive motor can be relatively small.

2.6.3   MODERN SINGLE-SCREW MACHINES

When the first thermoplastics materials such as cellulose acetate, plasticised PVC, polyethylene, etc., became available for extrusion almost the only screw machines in existence had been designed for the extrusion either of rubber or casein plastics. It was soon found that many of the design features which were good enough or even essential for rubber or casein extrusion were inadequate when used on the new thermoplastics. Gradually, therefore, since the early 1930s, a new class of single-screw extrusion machine has been developed specially for these materials (*Figure 2.2*).

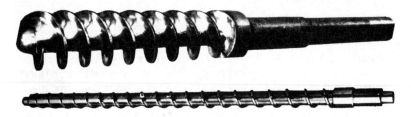

Figure 2.2.   Comparison of typical screws for rubber and for plastics. (Courtesy David Bridge & Co. Ltd)

A machine for the successful extrusion of thermoplastics materials must provide, in varying degrees according to the particular type of polymer being handled, the following features (which should be compared with the corresponding list in Section 2.6.2, giving requirements for rubber extrusion):

1 A feed opening suitable for the handling of cold material in granulated, chip, cube, or powder form, surmounted by a feed hopper.

2 A longer screw and barrel in proportion to their diameter, i.e. a greater L/D ratio. The barrel to be lined with or made from a corrosion-resistant alloy. Depending on the particular screw design, it will also give considerable heat due to frictional work and develop a pressure gradient along its length.

3 A breaker plate and screen pack and/or other restrictive system to be fitted.

4 A screw designed to provide compression, mixing and shear of the feed material.

5 The barrel to be heated to a high temperature in a number of zones, to give a heat gradient, and the temperature to be accurately controlled. Provision for controlled cooling of the barrel, to remove excess frictional heat, must also be considered.

6 A water-cooled feed opening.

7 Provision for water cooling or heating of the screw.

8 The feed screw to be easily interchangeable so that screws of different designs, both deeply and shallow cut, and with different compression ratios, can be used as required by the various polymers.

9 Provision for driving the feed screw at a wide range of speeds and for varying this speed without stopping the machine.

10 In order to stand up to the very high pressures which are often encountered in thermoplastics extrusion, machines for this purpose must be robust, capable of withstanding very high thrust loadings, and equipped with drive motors of ample power.

2.6.4 MULTI-SCREW EXTRUSION MACHINES

Apart from one brief reference the foregoing discussion has referred to normal extruders employing one screw only. Extruders employing two or more screws which may or may not have intermeshing flights are also available. The presence of further screws in the same barrel and the wide range of combinations and individual designs which may be used considerably influences the characteristics of the machine, although many of the fundamental requirements obviously remain unchanged. The design and operation of multi-screw machines is a complex matter, and is dealt with in Chapter 4.

2.7 Extrusion processes

Although this chapter has introduced some of the more important features of the extrusion machine—to be described in more detail in later chapters—it must be pointed out that the machine forms only one part of the extrusion process. Before passing on to treat the various features in greater detail, it is felt desirable, therefore, to stress the importance of other parts of the complete process.

When a thermoplastics material leaves the extrusion die it is usually in the form of a melt or a very soft mass. It often has little decisive form at this stage and even this would rapidly be lost unless it were handled in a suitable manner as it leaves the die. The product from the extrusion die, therefore, must, in a large number of cases, be looked upon as a semi-finished raw material which must be given its correct form and dimensions by subsequent processing whilst it is still mouldable[12]. The methods of doing this naturally depend on the thermoplastic being used and on the desired product; with the increasing tendency towards specialisation and the use of extrusion for the manufacture of one particular product only, more thought has been given and more time spent in designing the equipment for handling the extrudate. Although the operation has thus become more complex and less versatile it is also immeasurably more efficient.

2.8 Terminology

The variety of terminology which has grown up with increased knowledge and use of the extrusion process has often caused confusion. In an endeavour to standardise therefore, the Society of Plastics Engineers of the U.S.A. has drawn up an index of terms and definitions which is reproduced herewith with the courtesy of the above Society. The terms and definitions given in this index are used where possible in this book.

2.8.1 SCREW EXTRUDER

Screw Extruder—A machine which accepts solid particles or

molten feed, conveys it through a surrounding barrel by means of one or more rotating screws and pumps it under pressure through a forming orifice.

Extruder size—The nominal inside diameter of the extruder barrel.

Extruder bore—The inside diameter of the barrel.

Extruder length to diameter ratio, L/D—The distance from the *forward edge* of the feed opening to the forward end of the barrel bore divided by the bore diameter and expressed as a ratio where the diameter is reduced to 1, such as: 15 : 1 or 20:1. It is *not* based on the full flighted length of the screw.

2.8.2 EXTRUDER BARREL

Barrel—A cylindrical housing in which the screw rotates; including liner, if used.

Barrel liner or *Liner*—A removable sleeve in the barrel.

*Grooved liner (*or *Barrel)*—A liner whose bore is provided with axial grooves.

Rifled liner (or *Barrel)*—A liner whose bore is provided with helical grooves of a hand opposite to the hand of the screw flight.

Feed opening or *Feed throat*—A hole through the barrel for the introduction of material.

Vertical feed opening—An opening (round or rectangular) on the vertical centre line.

Side feed opening—An opening which feeds the material at an angle into the side of the screw.

Feed hopper—A funnel mounted directly on the feed throat, to hold a reserve of material.

Vent—An opening at an intermediate point in the extruder barrel to permit the removal of air and volatile matter from the material being processed.

Front barrel flange—A flange at the end of the barrel to which the die or adapting member is fastened.

Barrel heaters—The electrical resistance or induction heaters mounted on or around the barrel.

Barrel jacket—A jacket surrounding the outside of the barrel for circulation of a heat-transfer medium.

Heating zone—A portion of the barrel length having independent temperature control.

Zone temperature—The metal temperature of the barrel at the control point of the heating zone.

Control thermocouple—A thermocouple inserted in the metal wall to sense and control temperature.

Barrel coolers—Systems for the removal of heat, usually by air or liquid and normally controlled in zones corresponding to the heating zones.

2.8.3 EXTRUDER SCREW

Screw—A helically flighted shaft which when rotated mechanically works and advances the material being processed.

Screw flight—The helical metal thread of the screw.

Screw root or *stem*—The continuous central shaft, usually of cylindrical or conical shape.

Flight land—The surface at the radial extremity of the flight constituting the periphery or outside diameter of the screw.

Hardened flight land—A screw flight having its periphery made harder than the base metal by flame hardening, induction hardening, depositing of hard facing metal, etc.

Rear face of flight—The face of flight extending from root of screw to flight land on side of flight toward the feed opening. Same as 'Trailing Edge'.

Trailing edge of flight—The same as 'Rear face of flight'.

Front face of flight—The face of flight extending from root of screw to flight land on side of flight toward the discharge. Same as 'Pushing face of flight' or 'Leading edge'.

Leading edge of flight—The same as 'Front face of flight'.

Pushing face of flight—The same as 'Front face of flight'.

Bottom of screw channel—Surface of screw stem or root.

Bottom radius—The fillet between the flight and the root.

Screw shank—The rear protruding portion of screw to which the driving force is applied.

Screw shank seal—A sealing device to prevent leakage of material back around screw shank.

Torpedo—An unflighted cylindrical portion of the screw at the discharge end.

Mixing torpedo—Similar to torpedo but with a specially shaped surface.

2.8.4 DIMENSIONS OF SCREWS

Screw diameter—The diameter developed by the rotating flight land about the screw axis.

Root diameter—Diameter size of stem or root*.

Helix angle—The angle of the flight at its periphery relative to a plane transverse to the screw axis*.

Axial flight land width—The distance in an axial direction across one flight land.

Normal flight land width—The distance across one flight land in a direction perpendicular to the flight.

Flight lead—The distance in an axial direction from the centre of a flight at its outside diameter to the centre of the same flight one turn away*.

Flight pitch—Distance in an axial direction from the centre of a flight at its periphery to the centre of the next flight. In a single-flighted screw 'pitch' and 'lead' will be the same, but they will be different in a multiple-flighted screw*.

Screw flight depth—The distance in a radial direction from the periphery of the flight to the root*.

Diametral screw clearance—The difference in diameters between the screw and the barrel bore.

Radial screw clearance—The radial distance from the o.d. of the screw to the i.d. of the barrel bore. One-half the diametral screw clearance.

Full flighted length of screw—Overall axial length of flighted portion of screw including the tip and torpedo, if present.

Enclosed flighted screw length—The distance from the forward edge of the feed opening to the forward end of the screw flight, not including tips, torpedos, pressure cones, etc.

Number of turns of flight—Total number of turns of a single flight in an axial direction.

2.8.5   TYPES OF SCREWS

Single-flighted screw—A screw having a single helical flight.

Multiple-flighted screw—A screw having more than one helical flight such as: double flighted, double lead, double thread, or two starts, and triple flighted, etc.

Feed section (or *zone*) *of screw*—The portion of a screw which picks up the material at the feed opening (throat) plus an additional portion downstream. Many screws have an initial constant lead and depth section, all of which is considered the feed section.

Transition section (or *zone*) *of screw*—The portion of a screw

---

* The location of measurement should be specified.

between the feed section and metering section in which the flight depth decreases in the direction of discharge.

Conical transition section or *Conical tapered section* —A transition section in which the root increases uniformly in diameter so that it is of a conical shape.

Wrap-around transition section—A transition section in which the root is always parellel to the axis of the screw.

Metering section (or zone) of screw—A relatively shallow portion of the screw at the discharge end with a constant depth and lead, and usually at least three turns in length.

Metering-type screw—A screw which has a metering section.

Constant-lead screw or *Uniform pitch screw*—A screw with a flight of constant Helix Angle.

Constant taper screw—A screw of constant lead and a uniformly increasing root diameter over the full flighted length.

Decreasing lead screw—A screw in which the lead decreases over the full flighted length (usually of constant depth).

Cored screw—A screw with a hole in the centre of the root for circulation of a heat transfer medium.

Water-cooled screw—A cored screw suitable for the circulation of cooling water.

2.8.6 BARREL AND SCREW COMBINATIONS

Screw channel—With the screw in the barrel, the space bounded by the surfaces of flights, the root of the screw and the bore of the barrel. This is the space through which the stock is conveyed and pumped.

Screw-channel depth—The distance in a radial direction from the bore of the barrel to the root*.

Axial screw-channel width—The distance across the screw channel in an axial direction measured at the periphery of the flight*.

Normal screw-channel width—The distance across the screw channel in a direction perpendicular to the flight measured at the periphery of the flight*.

Axial area of screw channel—The cross-sectional area of the channel measured in a plane through the axis of the screw*.

Developed volume of screw channel—The volume developed by the Axial Area of Screw Channel in one revolution about the screw axis*.

---

* The location of measurement should be specified.

Enclosed volume of screw channel—The volume of screw channel starting from the forward edge of the feed opening to the discharge end of the channel.

Compression ratio—The factor obtained by dividing the developed volume of the screw channel at the feed opening by the developed volume of the last full flight prior to discharge. (Typical values range from 2 to 4, also expressed as a ratio 2 : 1 or 4 : 1.)

Channel depth ratio—The factor obtained by dividing the channel depth at the feed opening by the channel depth just prior to discharge. (In constant lead screws this value is close to but greater than the compression ratio.)

Screw efficiency (volumetric)—The volume of material discharged from the machine during one revolution of the screw divided by the developed volume of the last turn of the screw channel expressed as a percentage.

2.8.7 OTHER EXTRUDER PARTS

Extruder base or *stand*—The metal structure on which are mounted the basic extruder components such as: barrel, thrust housing, gear reducer.

Reduction gear or *Gear reducer*—The gear device used to reduce speed between drive motor and extruder screw. Supplementary speed reduction means may also be used, such as belts and sheaves, etc.

Drive—The entire electrical and mechanical system used to supply mechanical energy to the input shaft of gear reducer. This includes motor, constant or variable speed belt system, flexible couplings, starting equipment, etc.

Thrust bearing—The bearing used to absorb the thrust force exerted by the screw.

Thrust—The total force exerted by the screw on the thrust bearing. (For practical purposes equal to the extrusion pressure times the cross sectional area of the barrel.)

Thrust bearing capacity—The load in kilogrammes (pounds) that can be sustained under normal extrusion conditions, at the highest recommended screw speed of the machine for an average bearing life (B-10 rating—Bearing Manufacturers Association) of 50 000 h.

Breaker plate—A metal plate installed across the flow of the stock between the end of the screw and the die with openings through it such as holes or slots. Often used to support a screen pack.

Screens—A woven metal screen or equivalent device which is installed across the flow of stock between the tip of the screw and the die and supported by a breaker plate to strain out contaminants or to increase the pack pressure or both.

2.8.8 EXTRUSION PROCESS

Stock—The material while it is being processed in the extruder.

Stock temperature—The temperature of the stock. The location of measurement must be specified.

Stock thermocouple—A thermocouple is immersed in the plastic stream to sense the stock temperature.

Extruder temperature—The temperature of the material as it comes from the die.

Extrusion pressure—The hydraulic pressure generated in the stock at the tip of the screw.

Die pressure—The hydraulic pressure of the stock measured immediately behind the die.

Extruder output—The weight of material discharged from the extruder per unit time, usually expressed as kilogrammes per hour (pounds per hour).

Surging—A pronounced fluctuation in output.

Plastifying extrusion—An extrusion process in which solid material is fed to the extruder and plastifying to a melt takes place within the extruder.

Melt extrusion—An extrusion process in which molten material is fed to the extruder.

Autothermal extrusion—An extrusion process in which the only source of heat is that generated through mechanical working of the stock with the mechanical energy supplied by the drive.

Autogeneous extrusion—The same as 'Autothermal extrusion'.

Adiabatic extrusion—A term sometimes incorrectly used for 'Autothermal extrusion'.

Screw speed—Number of revolutions of screw per minute.

The British Standards Institution has also produced a glossary of the terms used in the plastics industry, BS 1755: 1951, which is presently undergoing revision. At the time of writing, the new section dealing with extrusion was in the final draft form for early publication[13].

REFERENCES

1   BURNESS, A., CANN, J.R. and IONS, R., 'Extrusion of Unplasticised Poly-
    vinyl Chloride', *Br. Plast. mould. Prod. Trader*, **21**, 244(1949)
2   WESTOVER, R.F., 'Continuous Flow Ram Type Extruder', *Mod. Plast.*, **40**,
    130 (1963)
3   WISSBRUM, K.F., 'Force Requirements in Forging of Crystalline Polymers',
    Polym. Engng Sci., **11**, 28 (1971)
4   ADAMSON, P.S., *Cellulose Plastics: Part II*, Plastics Monograph No. C.7,
    The Plastics Institute, London (1955)
5   STANNETT, V., *Cellulose Acetate Plastics*, Temple Press, London (1950)
6   LODGE, R.M., in *Fibres from Synthetic Polymers*, Ed. Hill, R., Elsevier, Lon-
    don (1953), p.363
7   PRESTON, J.M., in *Fibres from Synthetic Polymers*, Ed. Hill, R., Elsevier,
    London (1953), p.379
8   COOKSON, J.F., and TUNNICLIFF, F., 'The Extrusion of Thermoplas-
    tics', *Trans. Plast. Inst., Lond.*, **19**, 38 (1951)
9   FRIESEKE & HOEPFNER G.M.B.H., Erlanger-Bruck, W. Germany, Techni-
    cal Literature, extruder type SP150
10 ANON., 'New Extruder of Swiss Design', *Int. Plast. Engng*, **2**, 19 (1962)
11 GREGORY, R.B., and HESTON, E.E., 'QLT. A New Concept in Extrusion',
    Int. Plast. Engng, **5**, 305 (1965)
12 FISHER, E.G., *Plast. Prog. Lond.*, 181, (1951)
13 BRITISH STANDARDS INSTITUTION, 'Glossary of Terms Used in the
    Plastics Industry. Part 2: Manufacturing Processes; Section 25: Extrusion', to
    be published

3

The Modern
Single-screw Extrusion Machine

3.1 General

The earliest attempts to formulate a mathematical theory of screw pumps are due to Rowell and Finlayson[1], who in 1922, and again in 1928, published expressions for the output, the power requirements, and the efficiency of single-screw machines. Their work was based on the general theory of lubrication developed by Reynolds[2] and depended on the solution of a particular case of the differential equation:

$$\frac{\partial^2 v}{\partial x^2} + \frac{\partial^2 v}{\partial y^2} = \frac{1}{\eta}\left(\frac{dp}{dz}\right)$$

(3.1)

which was presented by Navier[3] in 1822.

This equation, in the form shown, was originally applied to the problem of flow and pressure drop in pipes and channels, and the well-known Poiseuille[4] expression for viscous flow in capillary tubes is in fact a particular solution for this equation. Rowell and Finlayson discussed, in effect, the problem of a groove filled with a viscous medium over the top of which a moving plane was able to slide with a velocity v. Since in this part of the problem there was no pressure along the axis of the groove, the z axis in equation 3.1, then $dp/dz = 0$ and the equation with which these workers dealt was

$$\frac{\partial^2 v}{\partial x^2} + \frac{\partial^2 v}{\partial y^2} = 0$$

(3.2)

The local velocity was obtained by solving this equation and the integration of the solution over the cross-section of the screw channel resulted in an expression for the discharge.

A similar technique was used to obtain the leakage flow across the top of the screw flights and this problem was treated as a pressure flow through an infinitely wide narrow slit.

In 1930, Decker, in Mitteilungen das Entwickelungsabteilung der firma Paul Troester, Hannover, gave a rather brief formula

$$Q = MnaN\mu$$

for the output of an extruder, where Q = output, M = volume of one flight, n = number of flights, a = the degree of filling, N = rev/min of screw, and μ = coefficient of friction. This formula takes no account whatsoever of the viscous material in the extruder and assumes that the material in the granular state controls the output of the machine. No allowance is made for differences in pressure caused by changing the die orifice; and the changes in material temperature, and hence consistency, have all to be accommodated in changes in μ, the coefficient of friction between the barrel wall and the material.

In 1946 Rogowsky[5], or Rigbi as he was later called, published a derivation of the extruder flow equation similar to that of Rowell and Finlayson but was apparently unaware of their work.

Further work on the same lines as that of Rowell and Finlayson was carried out in 1951 by Pigott[6], who described a series of experiments with oil and rubber on a single-screw extruder. An attempt was also made by Pigott to take into account the effect of shear rate on the viscosity of the material passing through the machine but the results obtained suffered from certain inaccuracies.

A paper published in 1951 by Grant and Walker[7] discussed velocity distributions in screw threads and illustrated these distributions with references to experimental work. The importance of transverse flow and pressure in mixing plastics materials in an extruder was indicated and references were made to the theoretical and practical volumetric efficiencies of screws.

Another paper published in 1947 by Eirich[8] considered the single-screw extruder as a mechanism for developing high pressures and an expression for the maximum obtainable pressure was derived.

A treatise on single-screw extruders was published in 1952 by Maillefer[9] in which differential equations for flow in a channel are solved to give output equations. An attempt is also made in this paper to deal with the problem of solid flow in the feed end of an extruder.

A most outstanding study of the single-screw machine is con-- tained in a collection of papers[10] published in 1953 by a group of workers in the Du Pont organisation, U.S.A. A critical review of the majority of the literature on extrusion was undertaken and a fairly comprehensive theory of melt extruders was propounded. This work gave expressions for the output and power requirements of melt extruders and compared the values predicted by these expressions with results obtained from a series of experiments.

A comparative criticism of the methods and results of Rowell and Finlayson, of the Du Pont workers, and of Maillefer, is contained in a paper by Meskat[11] published in 1955. The validity of the hypothesis propounded or implied in linearising the basic Navier-Stokes equations to arrive at equation 3.1 is questioned, and the principle of superposition is examined critically, and its limits of applicability are discussed. Perhaps the most pertinent criticism of the flow equation offered by Meskat concerns the validity of the assumption that plastics materials are Newtonian at any time at all during their passage through an extruder.

Papers by Spencer and Dillon[12] on the behaviour of polystyrene under high rates of shear show that this material at least is not Newtonian under these conditions and it is known that many of the other well-known thermoplastics materials depart similarly from the Newtonian viscosity relationship:

$$\eta = \frac{\tau}{dv/dx}$$

where τ is the shear stress in the material due to the velocity gradient in a viscous fluid.

A comprehensive survey and report of the theoretical equations for extruders has been published in two works by Squires *et al*[13] and Schenkel[14]. In addition to a consideration of the melt section of the extruder, the feed section has been investigated by Darnell and Mol[15] and also in Russia by Rakhmanov[15]. Proof that the theoretical approach has been reasonably accurate has been given by a number of workers[17,18].

It will be noticed that most of the earlier theoretical work dealt only with processes taking place in single-screw machines and that it was usually only the purely melt or the purely granular zones of these machines which were treated. Little information has been published about the interrelation of the two zones or the behaviour of materials in the transition stage between the solid and the viscous state.

Experiments by Street[19] using a multi-coloured feed have shown the flow of material in the screw channel and also experiments have been carried out by Du Pont using a glass-barrelled extruder and injecting coloured fluid into a corn syrup extrudate. Photographs of the transition of granules to a melt have been taken for an extruder operating under adiabatic conditions[20]. Underwood[21], on the other hand, disagrees with the theory that it is the metering section which controls the output and suggests that the equations developed for the analysis of this are not sufficiently accurate when applied in practice. In an attempt to analyse the effects of different screw geometry a segment screw was designed and a series of experiments planned under differing conditions to obtain the maximum information. Some of the results obtained from this work are given.

Recent work based on the use of computer techniques in the development of mathematical theories to explain the mechanism of screw extrusion, to predict outputs and to optimise extruder screw design has suggested modifications to the classical theories and to the usually accepted concepts of material movement along an extruder screw. Hitherto one of the most baffling problems has been the mechanism of melting, which has always been difficult to visualise and to explain.

A melting mechanism recently studied by several workers and formulated mathematically suggests that the solid polymer is initially compacted into a continuous block in the feed zone of the extruder and is then melted in a thin film on the surface of this block which is in contact with the heated barrel. This melt film is collected by the moving screw flights to serve as a heating medium for the remaining unmelted polymer[22-24].

Practical and theoretical work still remains to be carried out on the mechanisms operating in extruders but the many interesting and thoughtful papers now being published indicate that much is being done to investigate in a scientific manner the many problems associated with the extrusion process.

3.2 Screw design and basic design calculations

The movement of viscous material in the flights of a single-screw machine is envisaged as being composed of three distinct types of flow—they are:

Drag flow
Pressure back-flow
Leak flow

'Drag flow' takes place by virtue of the fact that the molten material in the screw channel adheres both to the fixed barrel, and to the rotating screw. A simplified picture of the velocity distribution engendered by this relative motion is shown in *Figure 3.1*. If this were the only type of flow taking place in the screw flight, the velocity profile would be approximately linear, and if the moving surface had a velocity V the mean forward velocity of the material would be $V/2$.

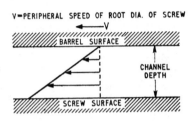

Figure 3.1. Drag flow

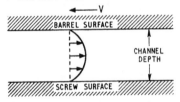

Figure 3.2. Pressure flow

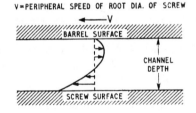

Figure 3.3. Net flow

'Pressure back-flow' arises when a restriction such as a die, valve or breaker place and screen is attached to the end of an extruder which gives rise to a pressure gradient in the channel. Pressure increases towards the die and the material is *assumed* to flow back down the screw channel although it has been shown[25] that there is in fact no *actual* movement of material in that direction. The velocity profile resulting from this is shown in *Figure 3.2*.

The net flow in the screw channel resulting from drag flow and pressure back-flow is determined by adding the two profiles algebraically and *Figure 3.3* indicates diagrammatically the resultant velocity, which is the flow of material in an axial direction in the barrel of the extruder and hence its output. The movement of any given particle of material is, however, of great complexity.

'Leak flow' is also caused by the pressure gradient along the screw and occurs between the lands of the thread flights and the barrel. The radial clearance between the lands and the barrel is normally very small, perhaps of the order of 0.12mm (0.005 in), and consequently the rate of flow in this case is much smaller than for drag flow and pressure flow. The effect of leak flow becomes more pronounced, however, as wear occurs either on the screw flight lands or in the barrel, and this has been analysed by Maddock[26] and others[27].

It is assumed that the algebraic addition of these three types of flow is justified and if the material is incompressible the total output of the extruder is given by the sum of the drag flow, the pressure flow, and the leak flow, i.e. $Q = Q_D - Q_P - Q_L$, since both Q_P (the pressure flow) *and* Q_L (the leak flow) will have opposite signs to Q_D (the drag flow).

3.2.1   FLOW THEORY FOR SCREW EXTRUDERS

Despite the apparent simplicity of the explanation given in the foregoing, a complete theory of the screw extruder which is valid for the whole range of thermoplastics materials is not likely to be attained. Such a theory would have regard not only to the dimensions, flight-pitch and other characteristics of the extruder, but would need also to take account of the physical properties of the material and how they change under varying working conditions.

But it is possible to idealise the extrusion process and by using a simple model to obtain by calculation a reasonable approximation to the performance of a screw extruder. The model usually taken is that of a screw pump operating on a viscous fluid in steady motion.

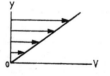

Figure 3.4

The most important characteristic of a plastics melt in calculations of this type is its coefficient of viscosity. Viscosity is a form of fluid friction. Suppose a fluid has a laminar motion, i.e. the molecules are moving in parallel planes, and suppose the velocity of any molecule to be proportional to its distance from a fixed parallel plane, say $v = ky$ as shown in *Figure 3.4*. This illustration, it will be observed, is similar to *Figure 3.1*, but is repeated here for convenience of reference.

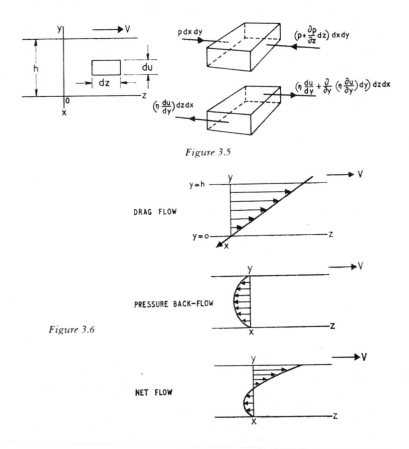

Figure 3.5

Figure 3.6

Then the distortion of any volume element of the fluid is a shearing strain. It is usual to assume that the shearing stress producing the shear is proportional to the rate of shear, and fluids in which this relationship holds are the Newtonian fluids referred to earlier in this chapter. This, however, is an empirical assumption and the subsequent analysis of fluid motion is not invalidated if some other law is substituted for it. If the fluid is not homogeneous, i.e. if, for example, the temperature at any point depends on the position of that point, then the value of η, the coefficient of viscosity, will also depend on that position, i.e. on the coordinates x, y, z (*Figure 3.5, 3.6*).

An example which can be used to provide a simple flow theory of a screw extruder can be given as follows:

A viscous incompressible fluid of infinite extent is in steady motion under pressure between two parallel plates at a distance h apart; one plate is fixed and the other moves in its own plane with constant velocity V. It is required to find the velocity at any point of the fluid. Let u be the velocity of an element of fluid at the point (x, y, z). Since the motion is steady, and ignoring extraneous forces, such as those due to inertia, the forces in the z direction arising from the pressure and from the viscosity must balance. These two forces are displayed in the diagrams on the right of *Figure 3.5*.

Therefore

$$\frac{\partial}{\partial y}\left(\eta \frac{\partial u}{\partial y}\right) = \frac{\partial p}{\partial z} \tag{3.3}$$

and since there is no motion parallel to $0x$, $dp/dx = 0$

it follows that $\partial p/\partial z$ is constant and if η is constant also

$$\eta \frac{d^2u}{dy^2} = \frac{\partial p}{\partial z} \tag{3.4}$$

Integrating this equation twice

$$u = \frac{1}{2\eta}\frac{\partial p}{\partial z}y^2 + Ay + B \tag{3.5}$$

If the viscous fluid is in contact with a rigid surface, then it is possible to give a plausible reason for the assumption that there is no relative motion of the fluid at that surface; for otherwise the frictional force at the surface must be very much less than that of the neighbouring filaments of fluid. Thus the constants A and B are such that $u = 0$ for $y = 0$ and $u = V$ for $y = h$. This gives

$$u = \frac{V_z}{h} - \frac{1}{2\eta}\frac{dp}{dz}(hz - z^2) \tag{3.6}$$

The Modern Single-screw Extrusion Machine

This is the difference between two terms; one due to the motion of the upper plate, namely the *drag flow* V_z/h, *and the other the*

pressure back-flow. $\dfrac{1}{2\eta}\dfrac{\partial p}{\partial z}(hz - z^2)$.

They are illustrated in the diagrams *Figures 3.1, 3.2* and *3.3* and again for convenience in *Figure 3.6*.

In giving the above example it has been assumed that the fluid extends infinitely in the *x* direction, in order to avoid complications due to further boundary conditions.

If the channel, of width w, of the screw is shallow, i.e. if h/w is small, then the above solution for the velocity distribution will be a fair approximation to the actual distribution. The boundaries are $x = 0$ and $x = w$.

The rate at which the extruder delivers the material, the cross section of the channel being rectangular, is

$$Q = w \int_0^h u \, dy$$

$$= \frac{Vwh}{2} - \frac{wh^3}{12\eta} \cdot \frac{dh}{dz} \tag{3.7}$$

A more elaborate solution avoiding some of the preceding approximations can be obtained by considering further the equations of steady motion of a viscous incompressible fluid as discussed briefly in the first section of this chapter. In vector notation, the Navier-Stokes equation is

$$V \nabla V = -\operatorname{grad}\left(\frac{p}{\rho} - u\right) + \frac{\mu}{\rho}\nabla^2 V$$

which is concerned again with laminar motion, the velocity u at the point (x, y, z) being in the direction of the z axis. Assuming again that the inertia terms and the extraneous forces can be ignored, the equation of motion reduced to equation 3.1, which is repeated here in slightly different form.

$$\eta\left(\frac{\partial u^2}{\partial x^2} + \frac{\partial x^2}{\partial y^2}\right) = \frac{\partial p}{\partial z}$$

This equation is, however, much more complex than the previous one, equation 3.7, and its solution, satisfying the appropriate boundary conditions, can be given different forms. There is only one solution, though, which satisfies the boundary conditions expressing the value of u at points of the boundary surfaces.

The problem, as before, is to find the steady motion of a viscous fluid in a channel of rectangular cross section height h, width w, the velocity of the boundary surface $y = h$, representing the barrel of the extruder, being V.

As before there is a drag flow and a pressure flow. The solution is therefore in two parts $v = v_P + v_D$. The *pressure flow* is

$$\eta\left(\frac{\partial^2 v_P}{\partial x^2} + \frac{\partial^2 v_P}{\partial y^2}\right) = \frac{\partial p}{\partial z}$$

with the boundary conditions

$$v_P = 0 \text{ for } x = 0 \text{ and } x = w$$

$$v_P = 0 \text{ for } y = 0 \text{ and } y = h \qquad \text{(the two flights)}$$

The *drag flow* is

$$\eta\left(\frac{\partial^2 v_D}{\partial x^2} + \frac{\partial^2 v_D}{\partial y^2}\right) = 0$$

with the boundary conditions

$$v_D = 0 \text{ for } x = 0 \text{ and } x = w$$

$$v_D = 0 \text{ for } y = 0 \text{ and } v_D = V \text{ for } y = h$$

Adding these two solutions

$$\eta\left(\frac{\partial^2 (v_P + v_D)}{\partial x^2} + \frac{\partial^2 (v_P + v_D)}{\partial y^2}\right)\frac{\partial p}{\partial z} \qquad (3.8)$$

with boundary conditions

$$v_P + v_D = 0 \text{ for } x = 0 \text{ and } x = \omega$$

$$v_P + v_D = 0 \text{ for } y = 0 \text{ and } v_P + v_D = V \text{ for } y = h$$

The solution is obtainable in the form of an infinite series

$$v_p = \frac{1}{\eta} \cdot \frac{\partial p}{\partial z} \left\{ \frac{y^2}{2} - \frac{hy}{2} + \frac{4h^2}{\pi^3} \sum_{n=0,1,3,5}^{\infty} \frac{1}{(2n + 1)^3} \times \right.$$

$$\left. \frac{\cos h \left[\frac{(2n + 1)\pi(2x - w)}{2h} \right]}{\cos h \left[\frac{(2n + 1)\pi w}{2h} \right]} \sin \frac{(2n + 1)\pi y}{h} \right\} \qquad (3.9)$$

$$v_D = \frac{4Uz}{\pi} \sum_{n=0,1,3,5}^{\infty} \frac{1}{2n + 1} \frac{\sin h \frac{(2n + 1)\pi y}{w}}{\sin h \frac{(2n + 1)\pi h}{w}} \sin (2n + 1) \frac{\pi x}{w} \qquad (3.10)$$

and the rate at which the volume of fluid is delivered is $Q_P + Q_D$,

$$Q_P = \int_0^w \int_0^h v_p \, dx \, dy \qquad \text{and} \qquad Q_D = \int_0^w \int_0^h v_D \, dx \, dy.$$

The expressions for Q_D and Q_P may also be conveniently expressed in a form due to Squires[25]

$$Q_D = \frac{nUzwh}{2} F_D \qquad (3.11)$$

where F_D is the 'shape factor for drag flow' and

$$F_D = \frac{16}{\pi^3 h/w} \sum_{n=0,1,3,5}^{\infty} \frac{1}{(2n + 1)^3} \tan h \left(\frac{(2n + 1)\pi h/w}{2} \right) \qquad (3.12)$$

$$Q_P = -\frac{mwh^3}{12\eta} \left(\frac{\partial p}{\partial z} \right) F_P \qquad (3.13)$$

where F_P is the 'pressure flow shape factor' and

$$F_P = 1 - \frac{192}{\pi^5} h/w \sum_{n=0,1,3}^{\infty} \frac{1}{(2n + 1)^5} \tan h \left[\frac{(2n + 1)\pi}{2h/w} \right] \qquad (3.14)$$

There still remains the application of this solution to the screw extruder. In a screw extruder, the barrel is stationary and the screw is in motion. But if the opposite of the angular velocity of the screw about its axis is superposed on the barrel, screw and plastics melt then a moving barrel with a stationary screw can be visualised. Because the inertia forces are small compared with those arising from the viscosity, the solution with stationary screw and moving barrel can be adopted and applied to the normal case of stationary barrel and moving screw.

It is also reasonable to expect the solution obtained to apply to the channel in a screw extruder ignoring the curvature distortion of the channel round the screw axis.

By substitution the combined drag and pressure flow equations can be reduced to

$$Q = F_D N \frac{\{\pi^2 D^2 h(1 - ne/t) \sin \phi \cos \phi\}}{2} - \frac{F_P \pi D h^3 (1 - ne/t) \sin^2 \phi}{12\eta} \frac{\partial p}{\partial l}$$

or
$$Q = F_D \alpha N - F_P \frac{\beta}{\eta} \frac{\partial p}{\partial l} \qquad (3.15)$$

where

$$\alpha = \frac{\pi^2 D^2 h(1 - ne/t) \sin \phi \cos \phi}{2} \quad \text{and} \quad \beta = \frac{\pi D h^3 (1 - ne/t) \sin^2 \phi}{12}$$

$$(3.16)$$

It will be noted that the first part of this equation, the drag flow, contains no factor for viscosity and is directly proportional to the screw speed N. Pressure flow as shown in the second part of the equation is influenced by viscosity η and channel depth h.

If we assume F_P and F_D to be unity, which is not unreasonable, then from equations 3.11, 3.13 and 3.16

$$Q = \frac{nUzwh}{2} - \frac{nwh^3}{12\eta} \left(\frac{\partial p}{\partial z} \right) \qquad (3.17)$$

or
$$Q = \alpha N - \frac{\beta}{\eta} \left(\frac{\partial p}{\partial l} \right) \qquad (3.18)$$

Mohr *et al.*[28] have, however, suggested a more convenient equation

$$Q = \frac{nUwh}{2}(1 - a) \qquad (3.19)$$

where a = ratio of pressure flow to drag flow.

From equation 3.15 for the special case of a single-start screw in which e is very small and can be neglected, substituting $t = \pi D \tan \phi$ gives the equation for Q as

$$Q = \frac{\pi^2 D^2 N h \sin\phi \cos\phi}{2} - \frac{\pi D h^3 \sin^2\phi}{12\eta}\left(\frac{\partial p}{\partial l}\right) \qquad (3.20)$$

It is now necessary to obtain a term for the leakage flow across the top of the screw flights in order to account for the third flow mechanism mentioned above as taking place in the extruder screw.

The leakage flow is considered as flow through a wide slit whose width is the length of one turn of the helix and whose thickness is the clearance between the outside diameter of the screw and the barrel wall. If pressure flow only is considered in this small clearance, the rate of flow is found by solving the equations for a fresh set of boundary conditions which results in a solution of the same form as the pressure term of equation 3.20. The solution can, therefore, be obtained quite simply from equation 3.20 if the dimensions for the slit are substituted for those of the screw channel.

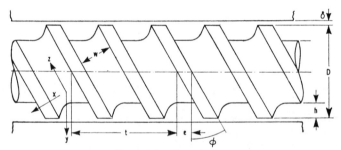

Figure 3.7a. Two-start screw

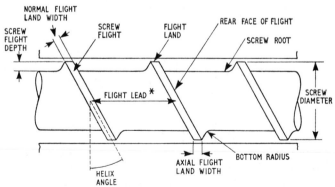

NORMAL FLIGHT LAND WIDTH

SCREW FLIGHT DEPTH

SCREW FLIGHT

SCREW FLIGHT

FLIGHT LAND

REAR FACE OF FLIGHT

SCREW ROOT

FLIGHT LEAD *

SCREW DIAMETER

HELIX ANGLE

AXIAL FLIGHT LAND WIDTH

BOTTOM RADIUS

✳ NOTE. FLIGHT LEAD AND FLIGHT PITCH ARE THE SAME IN SINGLE—FLIGHT SCREWS

Figure 3.7b. Single-start screw

From *Figure 3.7a* it may be seen that the dimensions of the slit through which the leak flow is taking place are

height $= \delta$ length $= c \cos \phi$ width $= \pi D/\cos \phi$

Making suitable substitutions and taking account the overall pressure drop along the screw (ΔP) and introducing a factor E to allow for eccentricity of the screw, the leakage flow becomes

$$Q_L = \frac{\pi^2 D^2 E \delta^3 \tan\phi \, \Delta P}{12 \eta e L} \qquad (3.21)$$

By adding equations 3.20 and 3.21 for drag flow, pressure flow and leak flow, the complete delivery of the extruder may be obtained. Thus

$$Q = \frac{\pi^2 D^2 N h \sin \phi \cos \phi}{2} - \frac{\pi D h^3 \sin^2 \phi \, \Delta P}{12 \eta L}$$

$$- \frac{\pi^2 D^2 E \delta^3 \tan \phi \, \Delta P}{12 \eta e L} \qquad (3.22)$$

The theoretical calculation of leakage flow for both pressure and drag effects has also been developed by Mohr, Mallouk and Booy[42], who considered flow across the flight lands in a direction at right angles to the axis of the screw. This has been considered in two sections: (a) the leakage flow over the channel, and (b) the leakage flow over the flight land. The result given for output is

$$Q = \frac{\pi^2 D^2 N h (1 - ne/t) \sin \phi \cos \phi}{2} (1 - a - J)$$

$$- \frac{n \delta^3 \pi D N \cos^2 \phi}{2 h^2} \left(\frac{\eta}{\eta_L} \right) \left[bc + a \left(\frac{\pi D}{n \tan \phi} + e \right) \right] \qquad (3.23)$$

where $J = \delta/h$ and $c = 1 - J$.

This represents the combined pressure and drag flow equations for output in the first term and the very much smaller leakage flow factor in the second term.

The delivery of the screw, given by equation 3.22, is the rate at which the material passes through the die of the machine, and the equation governing the flow of the material through the die is

$$Q = k\frac{\Delta P_D}{\eta} \qquad (3.24)$$

where k is a constant depending on the geometry of the die and ΔP_D is the pressure drop through the die. Equations 3.22 and 3.24 are, therefore, equal and ΔP is equal to ΔP_D.

It is now possible to obtain a value for the helix angle which gives the maximum output:

By eliminating ΔP between equations 3.22 and 3.24 and ignoring the very small leakage flow term an expression can be obtained for Q. This expression can be partially differentiated with respect to ϕ and h and equated to zero. If then the expressions obtained are solved for $\sin \phi$ and h respectively, and the two equations solved simultaneously, a value of 30° is obtained for ϕ.

$$Q = \frac{Ah \sin \phi \cos \phi}{1 + BL^3} \qquad (3.25)$$

with $$A = \frac{\pi^2 D^2 N}{2} + B = \frac{\pi D}{2}$$ and

and with $\phi = 30°$ $$h = \left[\frac{24kL}{\pi D(\eta D/\eta)}(1 - ne/t) \right]^{1/3} \qquad (3.26)$$

In considering the theory of flow in an extruder which has been discussed in the previous pages it must be understood that certain ideal conditions have been assumed to exist and this is justified in that it enables the results to be of practical use. It is worth noting, however, some of the variations from the ideal which do occur in practice. Non-uniform viscosity across the channel depth caused by screw and barrel heating or cooling is one variation and this has been resolved by Strub[29] and Mori *et al.*[30].

Channel curvature can be accounted for in the output equation by including a channel curvature factor which can be applied with little error to screws having a small helix angle of less than 20°. M.L. Booy of Du Pont has derived correction factors for both pressure and drag flow for angles of a larger and more practical size.

Partially full screw channels which occur particularly in vented barrels have a marked effect on the drag flow and output although it is interesting to note that the pumping capacity of a partially filled channel can be greater than that of a completely full one.

So-called edge effects caused by the sides of the screw channel can affect theoretical performance and these have also been studied by Squires[25].

It is often suggested also that the use of Newtonian flow behaviour for simplifying this output formula would give an appreciable error in the result and recent work carried out by Union Carbide, who have developed a mathematical model for power-law flow, has to some extent confirmed this. The results of this work indicate a significant difference from calculations based on Newtonian flow and relate much more closely to the results obtained from experimental data. By using this model and developing a computer programme therefrom it has been possible to design an extruder screw within specified parameters to give optimum outputs under given conditions[31].

3.2.2 GRAPHICAL REPRESENTATION

The output equation may be written in the form

$$Q = \alpha N - \frac{\beta}{\eta}\left(\frac{\partial p}{\partial l}\right) \qquad \text{where } \alpha \text{ and } \frac{\beta}{\eta} \text{ are constants.} \quad (3.27)$$

This is the equation of a straight line with a negative slope and if it is plotted on an output/pressure chart for various values of N (the screw speed), results in a series of parallel lines, as in *Figure 3.8*. These parallel lines are called 'screw characteristics' and show the relationship between the pressure generated at the end of the screw and the speed of the screw.

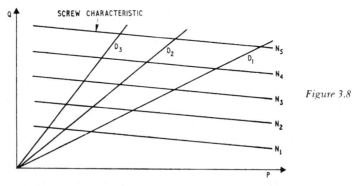

Figure 3.8

The slope of these lines is given by the terms for pressure flow and leak flow, where

$$\frac{\beta}{\eta} = \left[\frac{\pi D h^3 \sin^2 \phi}{12\eta L} + \frac{\pi^2 D^2 E \delta^3 \tan \phi}{12\eta e L}\right] \quad (3.28)$$

The second half of this term, which results from the leak flow, equation 3.21, is very small because of the dimension δ which makes δ^3 so small that terms containing it can be neglected without materially affecting the value of β/η. Hence

$$\frac{\beta}{\eta} \simeq \left[\frac{\pi D h^3 \sin^2 \phi}{12 \eta L} \right] \tag{3.29}$$

Thus the slope of the screw characteristic depends on the cube of the depth of the screw flight, on the square of the sine of the helix angle and is inversely proportional to the effective screw length. A small alteration in the flight depth will, therefore, have a considerable effect on the slope since the alteration would be cubed; the deeper the flight depth, the steeper the slope. Similarly, any alteration in the helix angle affects the slope of the screw characteristic in proportion to the square of the sine of the change in angle. An increase in helix angle produces an increase in slope.

On the other hand, since the slope is inversely proportional to the length of the screw, an increase in screw length would produce a flatter screw characteristic.

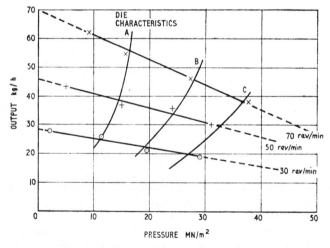

Figure 3.9. Diagram of a 60 mm extruder running at different speeds with different die characteristics. The screw had a short compression zone and polyethylene of melt index 0.5. The curved characteristics of the die plot are due to the reduced viscosity and increased throughput plus increased shear rate of material through the die

The series of straight lines through the origin $D_1 D_2 D_3$ ... in *Figure 3.8* represent the relationship between the pressure drop and the output through a die. They are lines representing Poise-uille's equation $Q = k \, \Delta P / \eta$, where k is their slope, and the dies increase in size from D_1 to D_3. The points at which these die lines intersect the screw characteristics are operating points for a particular die and screw speed.

By utilising these formulae the effect on output of different screw designs, die designs and other operating characteristics can be clearly seen. *Figure 3.9* shows the output and die pressure when operating at three different speeds with three different dies. This has been plotted from experimental figures.

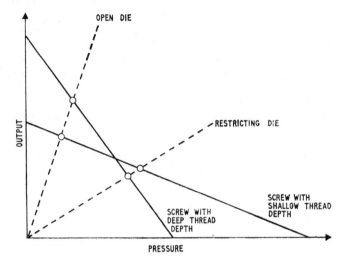

Figure 3.10. Combined effect of screw flight length and die design on output

A theoretical interpretation is shown in *Figure 3.10*. Here the effect of 'marrying' screw design to die design can be seen. The screw with the shallow thread depth will give a higher output from the restricting die and vice versa with the open die. A similar diagram *Figure 3.11* was developed by Schenkel with various other factors also plotted to indicate efficient operating areas for a single-screw extruder. By adjusting the operating conditions the intersection or working point can be moved to give the maximum results conducive to quality and other factors. For example, on the diagram the extruder is running at speed n, and the working point is A. To move point A the screw speed can be increased to n_2, which brings the working point to B where poor quality occurs. Increasing the die pressure rectifies this so that the working point is now at C. The shape and the position of the working area depend on the screw design.

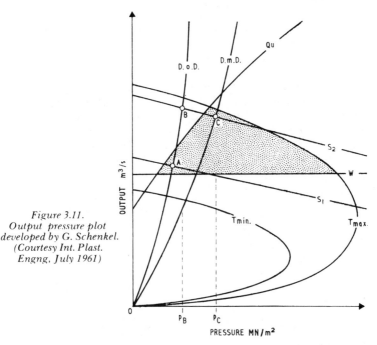

Figure 3.11.
Output pressure plot
developed by G. Schenkel.
(Courtesy Int. Plast.
Engng, July 1961)

A comparison of theoretical and measured working diagrams for different screw designs, die pressures, temperatures, screw speeds and materials has been made[17]. These compare favourably in most cases except in the higher speed range of a screw of progressively decreasing flight depth.

The output equation has been further developed by Ryder[32] to determine the pumping capacity in the second stage of a screw for a vented barrel extruder.

3.2.3 THE BEHAVIOUR OF GRANULAR MATERIALS

The feeding of granular material in an extruder can be treated to some extent as a problem of frictional forces exercised by the barrel and screw, which cause a solid nut of material to progress along the screw. Many materials in granular, flake, and even powder form, are compacted into a solid mass soon after entering the feed end of the screw and no shear or relative movement of layers takes place until the temperature of the material approaches its melting point. Therefore, during this initial stage, when the material is being heated as a solid, the classical laws of friction can be applied to the moving mass and a relationship obtained giving the pressure at any point along the screw and showing the effect of the thread profile on the mechanism.

The essential features of the problem can best be understood by referring to the simple case of a rectangular block of plastic material between two parallel planes, one of which is stationary whilst the other moves with a velocity V. A block of unit width moving with velocity Vm ($V > Vm > 0$) is shown diagrammatically in *Figure 3.12* and in the brief treatment which follows it is assumed that pressure inside the material is exercised uniformly in all directions.

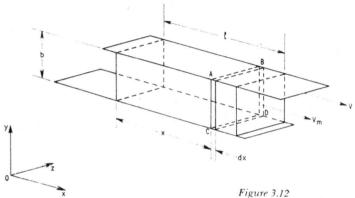

Figure 3.12

Consider a small element of material dx in width. If the pressure at the point x inside the material is p, then the force acting on the strips AB and CD is $p\,dx$. Therefore, if the coefficient of friction at AB is μ_s and that at CD is μ_v the classical friction relationship gives $p\mu_s\,dx$ and $p\mu_v\,dx$ as the forces in the x direction acting on the element at AB and CD respectively. The difference in these two forces gives the force pushing the element forward and is $p(\mu_v - \mu_s)\,dx$. Since this resultant force acts over an area equal to b the pressure pushing the element forward is $p(\mu_v - \mu_s)\,dx/b$ and this can be equated to dp, therefore

$$dp = p(\mu_v - \mu_s)\frac{dx}{b} \qquad (3.30)$$

From an examination of *Figure 3.12* it is apparent that μ_v must be greater than μ_s for the block to move.

The movement of material in the feed end of an extruder can be treated in a similar manner.

By considering the relative velocities of the material, the screw surface, and the barrel surface a relationship can be established between the output, the screw speed, and the thread profile. The relationship is of the form

$$Q = \pi DNwh \sin \phi \, (\cos \phi - \tan \alpha) \tag{3.31}$$

where Q is the output, D is the diameter of the screw, N is the rev/min of the screw, ϕ is the helix angle of the thread, and α is the angle between the material velocity vector and the normal to the relative velocity vector as shown in *Figure 3.13*.

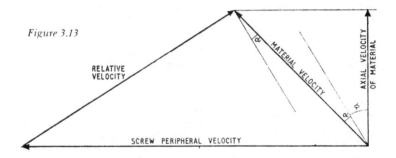

Figure 3.13

Furthermore, by considering the forces acting on the material an equation connecting the pressure gradient in the granular zone with the profile variables can be established, and is of the form

$$\frac{dp}{dx} = \frac{p}{h \sin \phi}[\mu_c \sin \alpha - K\mu_s] \tag{3.32}$$

where p is the pressure at the point, h is the depth of the thread, μ_c is the coefficient of friction between material and cylinder, μ_s is the coefficient of friction between material and screw, and K is a coefficient depending on the thread profile.

It is possible to eliminate α between equations 3.31 and 3.32 to form an expression which, when integrated, gives a relationship between the pressure and the output. This rather cumbersome relationship is of the form

$$\ln p = \left[\frac{\mu_c(A \cos \phi - Q)}{(A^2 - 2AQ \cos \phi + Q^2)^{\frac{1}{2}}} - K\mu_s\right]\frac{x}{h \sin \phi} \tag{3.33}$$

where $A = \pi DNwh$. However, $Q = Bw$, where B is a constant, and substituting this value for Q in equation 3.33 gives the expression for p

$$\ln p = \left[\frac{\mu_c(\pi Dwh \cos \phi - B)}{(\pi^2 D^2 w^2 h^2 - 2\pi DBwh \cos \phi + B^2)^{\frac{1}{2}}} - K\mu_s \right] \frac{x}{h \sin \phi} \quad (3.34)$$

which is independent of N (the speed of the screw (rev/min)).

The pressure profile in the granular zone is, therefore, an exponential curve, independent of output; a result which may possibly apply to a granular conveyor, but which cannot be said to hold for a plastics extruder.

From equation 3.32 it may be seen that the necessary condition for the machine to deliver is that $\mu_c \sin \alpha > K \mu_s$. This means that μ_c, the coefficient of friction between the material and the barrel wall, should be large and μ_s, the coefficient for the material and screw, should be small. The values of the respective coefficients of friction obviously increase with increase in temperature; thus in the feed section of the machine, where the material is heated by conducted heat, μ_c will generally be greater than μ_s. Furthermore, this condition is maintained by cooling the screw and its necessity can be appreciated when erratic feeding results from cutting off the screw cooling water.

The conveying rate for solids in the feed section of the screw has also been described mathematically in a classical work by Darnell and Mol[15] and can be calculated from the equation

$$\frac{Q_F}{N} = \frac{\pi^2 Dh(D - h) \tan \theta \tan \phi}{\tan \theta + \tan \phi} \quad (3.35)$$

To obtain θ the plug can be considered in static equilibrium so that the resultant of all the forces acting on it are zero, which gives the equation

$$\cos \theta = K \sin \theta A \quad (3.36)$$

where K is the ratio of forces acting perpendicular to reference plane to those acting parallel to it, and A is a collection of three terms from the full force balance equation derived by Darnell and Mol.

If the coefficient of friction of the material on the screw is zero the equation for θ reduced to

$$\cos \theta = K \sin \theta \quad \text{and} \quad K = \tan \phi$$

This simplifies the output equation to:

$$\frac{Q_F}{N} = \pi^2 Dh(D - h) \sin \phi \cos \phi \qquad (3.37)$$

and the drag flow capacity of the metering section is obtained from

$$\frac{Q_F}{N} = \tfrac{1}{2}\pi^2 (D - h)^2 \, h \sin \phi \cos \phi$$

This suggests that the volumetric delivery in the feed zone is about twice that in the metering zone. With bulk density varying between about 500 and 700 kg/m³ this would be satisfactory but as the friction between the screw and the material increases and back-pressure develops the balance is not maintained. Without back-pressure a compression ratio of at least 2.5 is necessary for polyethylene and even higher, 4.0, for rigid PVC and acrylics to prevent starving of the metering zone.

Maximum delivery from the feed zone of the screw is favoured by a deep channel, low friction between screw and material and a smaller helix angle than for the pumping of fluids.

If there were no friction between the screw and material, the helix angle for maximum output would be 45°. For materials such as polyethylene where $\mu = 0.25\text{-}0.40$ the helix angle is 22° and for acrylics where $\mu = 0.4\text{-}0.6$ it is approximately 17°. Experimental readings and calculated outputs have been compared for different screw and barrel finishes.

The output of the feed zone of an extruder has formed the subject of study in the U.S.S.R.[16], where Rakhmanov, after considering the work of Kruglikov and Rips, developed the equation

$$Q_F = \frac{\pi^2 D^3 \tan^3 \phi}{2} \left[2\cdot303 \log \frac{\sin \phi_1}{\sin \phi} + \tfrac{1}{2}(\cos^2 \phi - \cos^2 \phi_1) \right]_N$$

$$- \frac{\pi^2 D^3 \tan^3 \phi \tan p}{4}$$

$$\times \left[\frac{\sin (\phi - \phi_1)}{\sin \phi - \sin \phi_1} + \cos (\phi + \phi_1) \sin (\phi_1 - \phi) + (\phi_1 - \phi)\frac{\pi}{90} \right]_N$$

$$(3.38)$$

where ϕ is the helix angle at the flight land diameter $\tan^{-1} (t/\pi D)$, ϕ_1 is the helix angle at core diameter $\tan^{-1} \{t/[\pi(D - 2h)]\}$, and p is the angle due to the friction of the material against the body of the screw.

3.3   Variation of screw design with materials and die characteristics

Thermoplastics materials differ widely from each other in their mechanical and thermal properties. The surface hardnesses, the melt temperatures, and viscosities at melt, the coefficients of friction, the specific heats, and thermal conductivities of thermoplastics cover a wide range of values and since each of these factors influences screw design to some extent it is obvious that there must be many types of screw to deal adequately with these materials.

An extrusion screw should normally be designed with characteristics suitable for one particular material, since it is very rare that one screw can properly deal with a wide range of materials. In fact, the tendency is for a screw to be designed for a particular die/material combination, and more than one screw can be recommended for one material if the die setup is radically altered.

A typical screw is shown in *Figure 3.14a* and is divided into three zones: the feed zone, the compression or transition zone, and finally, the metering or melt zone. The screw in *Figure 3.14b* shows a special type of screw fitted with a torpedo or smear head, which imparts added shear to the material but no pumping action. This type of screw is not often used now, a longer-metering-zone design being generally preferred. *Figure 3.14c* is a design of screw with a very short compression zone which has been used with considerable success for the extrusion of high density polyethylene, nylon, and other hard materials of low melt viscosity.

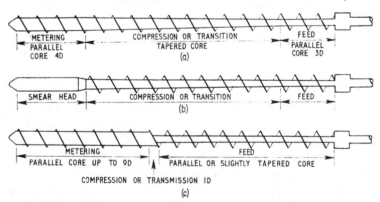

Figure 3.14a, b, c. Diagram showing three basic designs of screw

In the previous section a simplified theory of screw extrusion was developed which was based on the assumption that the material passing through the extruder was a liquid with a known viscosity coefficient. The theory in fact applied to melt rather than plastics extrusion. However, it is possible to apply the theory of a melt extruder to the metering section of an extruder screw and to obtain a reasonably accurate assessment of performance.

The theoretical design of extruder screws is now forming part of a three-year research programme at the University of Cambridge into polymer processing, sponsored by the Ministry of Technology. With the advance in knowledge of the behaviour of materials within the extruder and the change of state, the forces generated and absorbed and the physical state of the extruder as well as the melt it is now possible to write computer programmes for some sections of the screw and, as a result, predict the effect of changes in operating conditions, polymer or screw geometry[33].

3.3.1 FEED ZONE

The purpose of this zone as its name implies is to pick up the cold material from the hopper and to feed it to the compression zone. Feed materials, however, differ widely in their physical form and may be supplied as free-running fine powders, regular cubes, random cut chips with a percentage of fines, or even as small cylinders or spheres.

The feed section has to deal with materials whose form can be any one of those mentioned above and, as might be expected, it has been found by experiment that the helix angle most suitable for one form of material is not necessarily the best for another.

Also the coefficient of friction varies considerably according to the form of the feed material as well as its nature. The ideal helix angle for this zone, as previously explained, would be 45° with a hypothetical coefficient of friction between the screw and material equal to 0 ($\mu_s = 0$). But as the coefficients for most plastic materials used in extrusion are around 0.4 a helix angle of about 20° appears to be the most generally acceptable.

It has been shown previously that the most efficient helix angle for the metering zone of an extrusion screw is 30°, with a flight pitch of 1.8D, but since the majority of screws are of the constant pitch variety the usual compromise helix angle for the whole screw is normally based on that of the feed zone. Helix angles as great as 25° and as low as 10° are known but in practice 17.7° is most commonly employed, thus giving a pitch equal to diameter. The performance of the feed zone of the screw has a marked influence on the output of a machine but the influence of the helix angle in the other zones of the screw has a smaller effect[34].

As it is assumed that the metering zone of the screw controls the output it is important that the feed zone should be capable of conveying sufficient material to keep the metering zone full. On the other hand it is equally important to ensure that the supply of material from the feed zone is not too great to overrun the metering zone. A pronounced departure from this balance either way will result in surging or pulsation so that it is necessary to exercise reasonable care in the selection of a compression ratio to suit the bulk factor of the feed material. Compression ratios as normally used vary between about 1.5 : 1 and 4 : 1 and reasons for these figures and calculations to prove their validity have been given[13],[14].

Tests have also been carried out to determine the influence of the thread form in the feed zone of the screw[35] and, as might be expected, an angle of 90° between the root of the thread and the thrust face of the land gives the best performance.

It must be noted however that such a thread form when used in the metering and compression sections could result in 'dead spots' and consequent degradation of material. It is therefore common practice to design screws with radiused fillets at the flight roots which besides giving greater strength are more readily cleared of material at each revolution of the screw.

As mentioned in the theoretical section, it is necessary that the frictional properties between the material and the bore, and between the material and the screw in the feed section, should bear the correct relationship to each other. This relationship is maintained by cooling water in the screw and for some rigid PVC materials, and powdered materials generally, grooves have been cut in the barrel bore at the feed zone to increase the friction between the material and the barrel.

3.3.2　COMPRESSION ZONE

The compression zone, or transition zone as it is perhaps more accurately termed, follows immediately after the feed zone and can be formed by the gradual increase of the root diameter of the screw thread until the diameter of the metering section is reached. Shorter compression zones, which often comprise two or even one or less pitches of the thread only, represent an alternative design which has been found to operate very well with most materials except perhaps the highly viscous heat-sensitive polymers such as rigid PVC.

The compression zone must be designed not only to compact the material but also to conform to the rate of melt and the volume change as the material passes from the solid to the viscous state. It should aim at introducing the correct amount of compacting, firstly to push any occluded air back to the feed zone, and secondly to compact the material to improve its thermal conductivity. Furthermore, during its passage through the compression zone the material should become sufficiently viscous and deformable to be able to absorb energy from shear so that it may be heated and mixed uniformly through its mass. Thus, by the time the material has passed through the compression zone, it should be homogeneously melted and devoid of nibs or unmelted particles so that the next zone has only to deal with a viscous fluid.

The compression zone of a screw is probably the most difficult mechanism to define and is the one part of an extruder which has not so far formed the subject of detailed mathematical study. The behaviour of the plastic during its residence within this zone has been studied, however, by several workers[19],[20] using transparent barrels but although a pattern of behaviour has been observed it has not yet been possible to develop this into a general theory to be of benefit in the design of extruder screws. The profound complexity of this zone results from the gradual transition from solid to laminar flow and from the increasing effect of shear. The melt viscosity and heat transfer both change within a short or long length of screw depending on the melt characteristics of the material and the feed stock at this stage consists of a continually changing suspension of solid particles in a molten matrix. The position of the transition point along the length of the screw is believed to have a considerable influence on the quality of the extrudate.

As was stated previously, the most common method of achieving a compression ratio on an extruder screw is to decrease the depth of flight over a certain distance and so effect either a gradual or a rapid reduction in the cross-sectional area of the screw flight. However, there is another method of achieving a compression ratio which is considered preferable for certain applications. This design consists in gradually reducing the pitch of the screw whilst maintaining the depth of flight constant. By this means, the rate of shear in the material, which is inversely proportional to the depth of flight, is maintained constant and the energy input to the material is less than would be the case for a decreasing flight depth system. Since in most materials, however, the quality of the extrudate improves with increasing shear the use of this design. is limited to heat- or shear-sensitive polymers.

3.3.3   METERING ZONE

The metering zone is the final part of the screw and acts rather as a metering pump from which the molten plastics material is delivered to the die system at constant volume and pressure. The ratio of the volume of one screw flight in the feed zone to the volume of a flight in the metering zone is termed the compression ratio of a screw.

More is known about the behaviour of this part of the screw than any other, since the material in it is completely viscous and the equations discussed earlier can be applied approximately to the processes taking place therein.

The mechanisms of drag flow, pressure flow and leak flow may be envisaged as operating in the metering zone; and the interaction of the variables of this part of the screw and die system can be studied to some extent mathematically and verified by experiment. In *Figure 3.15* a typical extruder screw output curve is shown, plotted against pressure for a variety of die sizes and a constant screw speed.

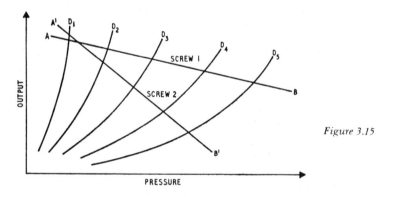

Figure 3.15

When considering the pressure developed by an extruder screw in the metering zone it is generally assumed that pressure at the die represents the pressure increase over the metering zone, but any pressure at the commencement of the metering zone will increase this figure, and in consequence the calculated output, according to the diagrams and equations given in the previous sections will be higher than the actual output. Maddock[36] has shown that the pressure at the start of the metering section can exceed the pressure at the die with certain screw designs. The result of this is that outputs are higher than calculated due to a forward pressure flow.

Screw speed change does not alter the basic shape of the graph, *Figure 3.16*, with the type of screw used in these experiments which had a 22:1 *L/D* ratio with a four-turn metering zone and a sixteen-turn transition zone of the increasing core diameter type.

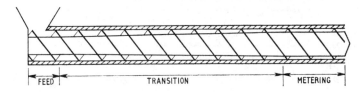

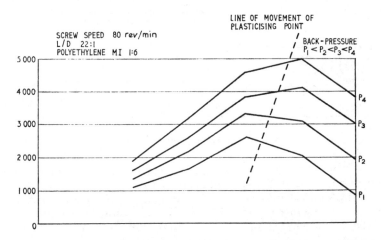

Figure 3.16. Diagram showing movement of maximum pressure point with increase of pressure

Further investigations showed that this pressure peak and pressure drop could be overcome by increasing the metering zone length and decreasing the transition length. A better mixing and higher output could also be obtained. The channel depth of the metering zone needed to be increased however to prevent over-heating of the material. No alteration to the basic design of the transition zone was made and it remained an increasing core diameter type (*Figure 3.17*). As a result of later work it has been further suggested by Maddock that a mixing section could with advantage be incorporated in the screw to split the metering zone into two parts. Such a mixing section would ensure complete melting and homogeneity by forcing the material through a thin slot either over an additional screw flight or by means of a fluted section[37].

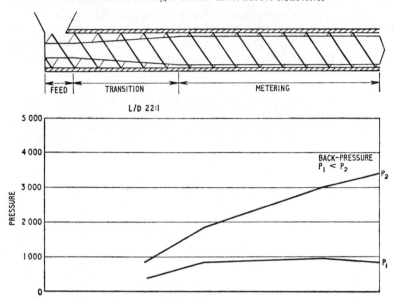

Figure 3.17. Diagram showing the effect of longer metering zone on pressure peak

A detailed study, both theoretical with the aid of computers, and experimental, on the metering zone performance has been carried out by Marshall and Klein *et al.*, the results of which were given in papers presented at the S.P.E. Annual Technical Conference, March 1965, and subsequently published in the *S.P.E. Journal* in October, November and December of that year.

Schiedrum and Domininghaus[17], in investigating screw characteristics, suggested that the overlapping of processes within an increasing-core-diameter screw is an important cause of inconstant delivery. The movement of the peak pressure point (*see Figure 3.16*) will have an immediate effect on output and quality. Their suggestion is that the feed zone of the screw should have a moderately deep channel, constant pitch and constant core diameter but should extend for some four fifths of the overall screw length. This is in effect a conveying screw and barrel heat conduction provides the largest percentage of heat for melting the material. The transition zone is short—approximately *D* —and the metering zone about a fifth of the length. In the transition zone a very severe increase in shear takes place and this type of screw would not therefore be satisfactory for rigid PVC or similarly viscous materials owing to the risk of local overheating. *Figure 3.18* shows the pressure increase for a short-transition-zone screw of the type suggested and it can be seen that the pressure peak is near the end of the screw so that the risk of pulsation is much reduced.

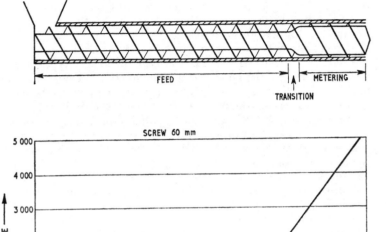

Figure 3.18. Diagram showing the pressure gradient in a short-transmission-zone screw

The type of plot shown in *Figure 3.15* differs only slightly from the theoretical diagram on which it is based. The main difference is that the die lines $D_1 \ldots D_5$, which are the pressure output relationships for a variety of dies whose cross-sectional areas decrease from D_1 to D_5, are curved when plotted from actual recordings, whereas in theory they are straight lines through the origin. This deviation is a characteristic of non-Newtonian fluids, such as molten thermoplastic materials.

The points at which the screw characteristics AB and A′B′ cut the die characteristics represent the pressure at the screw tip and the output of the machine for a particular die at a given screw speed.

Confining attention to the line AB, it is possible to read off from the diagram the output which screw 1 would give at the given screw speed with any of the dies. If now screw 1 were replaced by a screw with a deeper metering zone, or, in other words, by a screw with a lower compression ratio, the resulting output/pressure relationship would be given by a line with a steeper slope. This means that screw 2 would give a much lower output on a small die but that the difference in output between the two screws would diminish as the die size increased until on a large die it would be possible to obtain a higher output with the deep flighted screw.

A diagram of the type shown in *Figure 3.15* and in *Figures 3.8, 3.9* and *3.10* in Section 3.2.2 make it possible to specify the screw required for a particular die arrangement. If, for example, it is required to feed a narrow slit die, or a die system in which there is a great deal of restriction, a screw with a relatively high compression ratio would be specified. On the other hand, a large die offering little resistance to flow would be best coupled with a lower compression ratio screw. Usually, of course, it is necessary to strike a compromise and to specify a screw to give a fairly consistent output over a wide range of dies, in which case the higher compression ratio would be specified because it is relatively unaffected by die size.

The desirable depth of metering zone on a screw is very closely related to the mean viscosity of the molten plastics material passing through the section and cannot be accurately specified until the working viscosity of the melt is known. Thus a screw whose melt zone results in a flat characteristic with polyethylene at normal extrusion temperatures would produce an extremely steep characteristic with hard vinyl at working temperatures. Similarly, a screw designed for unplasticised PVC material would result in a very steep characteristic if operated with polyethylene.

3.3.4   THE SMEAR HEAD OR TORPEDO

In *Figure 3.14b* the metering zone has been replaced by a plain parallel section whose shape is roughly that of a torpedo, from which it often takes its name. The clearance between the diameter of the smear head and the barrel bore is usually less than the depth of the screw flight which immediately precedes this section and, although the surface is usually smooth and parallel, it may be fluted, knurled, or tapered, and may even have a special thread on it.

There are three purposes which the smear head serves: firstly, it acts as a mixer; secondly, it enables the material to be heated by mechanical working and thirdly, it forms a special type of restrictor. The mixing is carried out by virtue of the fact that the clearance between the smear head and the barrel wall is very small, and in consequence a high shear rate is set up. In addition to this, the various fluting and knurled surfaces, which many smear heads have, serve to break up and mix the material passing through this zone. The second purpose of the torpedo is almost identical with the first, inasmuch as the small radial clearances which promote good mixing inevitably result in a high energy input to the material from the screw, because of the high rate of shear which they engender.

The third reason given for specifying a torpedo was to build up a back-pressure by virtue of the restriction to flow which it occasions. Since pressure is only built up by the smear head, the pressure profile along the screw shows a drop across the torpedo. The high pressures produced just prior to this section are considered to assist mixing and, unlike a screen pack system, the smear head restrictor is self-cleaning.

3.4 Types of screw

The range of plastics materials available to the fabricator is very wide and, as mentioned at the beginning of Section 3.3, it is usually necessary to design a special screw to cater with maximum efficiency for the physical and thermal properties of each individual material. In addition to this, a case can often be made for designing a screw to suit a particular material/die combination so that more than one screw could reasonably be specified for a given material. Under these circumstances, a detailed survey of screw usage would entail a far more comprehensive examination than can be embarked upon in this work.

The most common screw designs adopted at the present time are the increasing-core-diameter type and the short-transition-zone type, the former being more versatile in its application. The torpedo and smear head have largely been disregarded in favour of an increased length of metering section. A limitation on the length of small-diameter screws, say up to 60 mm (2½ in) in diameter, is imposed by the torque load in which the screw can take. Screws of up to 35 : 1 L/D ratio are now in common use and screws of up to 40 : 1 are known for special applications.

With the adoption of the two-stage screw and venting described in Section 3.9 the length has had to be increased, and 27-30D is not uncommon. The advantages of venting are, however, considerable particularly with materials such as the cellulosics, acrylics and polyamides.

The increasing-core-diameter screw provides a good general design for a large variety of materials giving good heat transfer characteristics and controlled shear and, providing the metering zone has sufficient length, a uniform output. The long transition zone can however cause a hold-up in material flow and unequal movement in the melt. The amount of shear imparted is also less than with the short transition zone and consequently the mixing characteristic of this screw is not so high. For heat-sensitive materials such as rigid PVC this is an advantage.

With the short transition screw the rapid change from feed to metering zone creates a large amount of shear on the material and can cause local overheating particularly with materials of high viscosity. With materials of low viscosity and relatively sharp melting points such as nylon, the polyolefins and polyvinylidene fluoride for example, screws of this design give very good results.

Most if not all single-screw extruder screws are bored for circulating water cooling or oil heating systems, and broadly speaking the specification of screw design depends on the melt temperature range, the melt viscosity and the stability of the thermoplastic material.

Different workers have studied the design of screws for different materials. Pfluger[38] and Toll[39] have evaluated different screws for the extrusion of nylon and Garber and Cassidy have studied screws for acrylic extrusion[40]. A general comparison of theoretical screw design and practical operation has been given by Kreft and Doboczky[41], who maintain that the screw geometry is the key to quality output. These workers also suggest that theory and practice are sufficiently close for practical purposes, although it may be found that better results are obtained by the use of shallower channel depths and higher screw speeds than is suggested by theory.

Some of the newer materials now being extruded present special problems, and these are discussed in later sections of this volume.

3.5 Heat and power requirements

One of the main functions of a plastics extruder is to raise the temperature of the feed material usually from ambient to a temperature at which it can flow and be suitably formed to a desired shape. It is, therefore, necessary to supply heat energy to the material at a rate which will produce the temperatures required for this operation. In addition to the energy necessary to heat the material, power must be supplied to the screw in order to turn it against the frictional and viscous resistances of the material in the screw flights.

Electric resistance or induction heaters are the usual methods of supplying heat energy to the plastics material by conduction and the screw is driven by an electric motor or by an hydraulic motor.

It will be readily appreciated that these two energy mechanisms are closely interrelated since an alteration in the rate of supply of one will produce a change in the other. For example, the higher the temperature to which the heating system raises the plastics material, the lower the energy required to turn the screw. Furthermore since the energy supplied to the extruder screw reappears as heat in the material, the greater the energy supplied to the material by the screw, the lower the energy requirements from the heaters.

So far, no mathematical relationship has been derived to describe completely the interaction of these two processes but a simple theory has been developed to describe the power requirements of an identical melt extruder. This theoretical work is a further development of the flow equations discussed earlier.

The power absorbed by the screw can be derived from

$$dz = \psi_s \, dM_s + dZp + \psi_L \, dM_L$$

where $\psi_s \, dM_s$ is the power absorbed by viscous shear energy in the screw channel, dZp is the power required for pressure, $\psi_L \, dM_L$ is the power absorbed by shear energy between the flight land and the barrel, and ψ is the dissipation function for viscous shear energy. Each of these terms of the equation can be found from established data so that the total screw power requirements can be formulated. Mohr *et al.*[42] and Gore and McKelvey[43] have developed such formulae.

$$\psi_s \, dM_s = \frac{\pi^3 D^3 N^2 \eta (1 - ne/t)}{h}[(1 + 3a^2)\cos^2 \phi + 4\sin^2 \phi] \, dl \quad (3.39)$$

$$dZp = Q \, dp \quad (3.40)$$

$$\psi_L \, dM_L = \frac{n\pi^2 D^2 N^2 \eta_L l}{\delta \tan \phi} \, dl \quad (3.41)$$

giving a formula for total power requirements of

$$dZ = \frac{\pi^3 D^3 N^2 \eta (1 - ne/t)}{h}(1 + 3\sin^2 \phi) \, dl + Q \, dp$$
$$+ \frac{n\pi^2 D^2 N^2 \eta_L l}{\delta \tan \phi} \, dl \quad (3.42)$$

where $l = 2 \sin \phi$

A second method of calculating the total power requirements of a screw can be derived from the viscosity and shear rate at the barrel wall, the area and the screw speed.

The first term, which is proportional to the square of the screw speed, shows the power dissipated in shearing the material in the screw flights. The second term gives the power required to maintain the pressure in the screw and the final term, also proportional to the square of the screw speed, shows the power used in shearing the thin film across the top of the screw lands.

The same limitations apply to the use of this power relationship as those outlined in Section 3.2 for the output equation. That is to say, the equation refers only to a melt extruder through which a Newtonian melt is passing and can, therefore, only give an approximate picture of the power requirements of the melt zone of a plastics extruder. However, the equation does at least indicate the effect which alterations in any of the screw variables would have on the power requirement.

For example: the last term in equation 3.42 is the largest; thus, a slight variation in the clearance between the screw and the liner results in a relatively large change in the power requirement. From the first term it may be seen that the deeper the flight of the screw the smaller is the amount of power needed to turn the screw; thus a high compression ratio screw would need a larger motor than would a screw of low compression ratio.

In practice the amount of energy required by the screw is also an important factor in the economics of extrusion. It is not, however, possible to control this by design only as it is a function of speed, die design and material viscosity. The energy absorption of the system can be assessed by

$$Z_s + G(Z_m - Z_1)$$

where Z_s is the power input to the screw, Z_m is the power given by the motor, G is the efficiency of the motor, and Z_1 is the power under 'no-load' conditions. It is more important to relate the power input to production capacity giving a figure of kilowatts per hour, per kg, or lb. Schiedrum and Domininghaus[17] give several examples of this with different types of screw. The maximum value for high density polyethylene for example is

$$\frac{Z_m}{Q} = 1 \cdot 08 \text{ MJ/kg}$$

where Q is the output. Plotted curves show clearly that the lower the die resistance the lower the specific power requirements; similarly, as temperature increases so the power input decreases. With polypropylene there is a much greater increase in power consumption per unit of output than with polyethylene or plasticised PVC.

The power absorbed by the screw can be considered to be used for applying shear to the material and for applying pressure to it. If the shearing energy, which is converted into heat, is equal to or greater than the enthalpy of the material, the machine operates adiabatically. With increasing core diameter screws it has been found that shearing energy required for polypropylene is higher than with the short-compression-type screw in all cases except for very low speeds.

From the power absorbed by the screw Z_s, and the screw speed, the torque can be calculated. From this and the knowledge of the permitted torque on the particular size of screw it is a reasonably simple matter to calculate the permissible length of screw for specified speeds or the maximum speed for any particular screw design and length.

3.5.1 SCALE-UP

The power equation also gives a very useful guide to the relationship between the size of an extruder and the power requirement. In the first term of the equation, the screw diameter D appears cubed; thus, the scale-up rule relating to extruders of various sizes shows that in theory the motor size should be proportional to the cube of screw diameter if all other screw dimensions are changed in proportion to the power indicated in equation 3.42.

As stated above, the equation can only accurately refer to a melt extruder pumping a Newtonian fluid whose mean viscosity η is known, and these conditions are never satisfied in a plastics extruder. However, experiments have shown[44] that for a constant material-output temperature the power requirement of a plastics extruder is roughly proportional to the square of the screw speed. From this it has been argued[45] that there is a limit to the speed at which a single-screw machine may be operated. Further experiments[50], however, have failed to substantiate this over a large range of screw speeds. Support for the experimental data has been given in that the rate of heat transfer increases as the square of the barrel area and the pumping capacity of the screw turning at the same peripheral velocity increases as the square of its scale-up factor. The square of the channel size is proportional to the necessary residence time for homogeneity and quality. In practice, therefore, the larger machine has a reduced screw speed and often a proportionally shallower channel giving a capacity below the theoretical figure. In modern extruders, therefore, the power requirements are adequately covered.

A detailed analysis of scaling-up and practical experiments supporting the theory have been given by Carley and McKelvey[47]. For optimum design in which back flow is about a third of the drag flow, Carley suggests[48] the use of a simplified equation for calculating the power requirement

$$hp = 6 \times 10^{-4} Q \, \Delta p$$

3.6   Residence time

In order to increase output on a machine with a given screw diameter it is necessary to increase the screw rotational speed. As a result of this it may be necessary to increase also the length of the screw to obtain the minimum residence time for the material to reach complete homogenisation. This, of course, in its turn will increase the power required and the torque.

Residence time can be calculated from the equation

$$R = \frac{M}{Q}$$

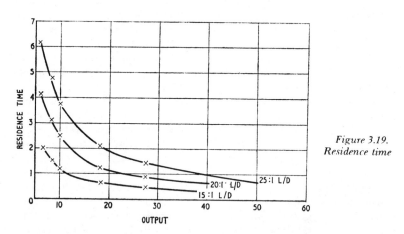

Figure 3.19.
Residence time

As the bulk density of the material changes during its passage to the transition point in the extruder this must be taken into account and also the specific gravity of the material at different temperatures. The calculations show that the required residence time with a screw of increasing root diameter is less than for a screw with a short compression zone. A comparison of residence times against output for three screws of the same diameter but different lengths is shown in *Figure 3.19*.

3.7 Surging or pulsating

The question of pulsation is perhaps more relevant to a chapter dealing with operational techniques than with screw design, but as the effect can be minimised by correct design it is pertinent to consider its characteristics here.

In an attempt to analyse the pulsating effect, Kirby[49] has developed an equation based on the theory of kinematic flow and storage. Starting from the basic output equation for adiabatic operation, an equation for surging which describes the stability of the system for any degree of disturbance is obtained:

$$\frac{1}{\Delta Q}\frac{dQ_0(t)}{dt} = \frac{1}{A}\frac{K\alpha BN}{(KL + B)^2} \tag{3.43}$$

l = screw-lead = $\tan D$
t = time
L = length of flight filled with melt
K = die constant = $\dfrac{1}{\text{die pressure}}$

$$\alpha = \frac{\pi^2 D^2 h}{2}\left(1 - \frac{e}{l}\right)\sin\phi\cos\phi \qquad e = \text{flight width}$$

$$B = \frac{\pi Dh^3}{12}\left(1 - \frac{e}{l}\right)\sin^2\phi \qquad Q_0(t) = \frac{KL\alpha N}{KL + B} = \begin{array}{l}\text{output at}\\ \text{any given}\\ \text{time}\end{array}$$

N =   screw speed

A =   cross-sectional area of channel section filled with melt

From this it is possible to express the equation in terms of K, α, N, B, A and Q_1 where $Q_1(t)$ is the rate at which melt enters the filled channel, $t = 0$

$$\frac{1}{\Delta Q}\frac{dQ_0(t)}{dt}\Bigg]_{t=0} = \frac{K}{\alpha NBA}(\alpha N - Q_1(t))^2\Bigg]_{t\leq 0} \tag{3.44}$$

An increase in die pressure (decrease in K) and an increase in $Q_1(t)$ will decrease the surging figure and increase the melt channel length. From this a basis for screw design can be obtained. Helix angle can be decreased so that conveying rates approach drag flow or a longer screw can be used. Whatever method is employed it is the result which is important and in this case it is the length of screw channel filled with melt which gives stability to the system. This has been confirmed by Krueger[50] in experiments with two screws having a constant channel depth in the feed zone, a short compression zone and a metering zone. Using a glass cylinder the actual melt length was observed and plotted under varying conditions.

3.8   Adiabatic extrusion

At the beginning of the last section the two mechanisms contributing energy to the extrusion process were discussed, and mention was made of the fact that the energy supplied to the screw reappears as heat in the plastics material in the flights. It is, therefore, possible to use the extruder screw as a heating device; and if sufficient power is available to raise the material temperature to the level required for satisfactory extrusion, and no heat is contributed by external heaters, then the system is said to be adiabatic. In the strictest sense of the term the system is not, in practice, adiabatic, since heat leaves the system both by radiation and in the screw cooling water. A more descriptive term would be exothermic extrusion, meaning literally an extrusion process wherein heat is generated. The term 'adiabatic extrusion', however, is the more widely used in this connection and the process will, therefore, be termed adiabatic in the discussion which follows.

The output equation 3.22 in Section 3.2 shows that the output of an extruder is directly proportional to the screw speed, and equation 3.42 of Section 3.5 shows that the power input to an extruder from the screw is proportional to the square of the screw speed. Therefore, as the screw speed increases the energy supply per kilogramme of material increases until, at a certain screw speed, all the required energy is supplied to the material by the screw. At this speed, and above, the system is adiabatic. Although it has been stated in previous sections that the output and power equations can only be partially applied in practice, it is nevertheless possible to work a machine adiabatically although it is not yet possible to calculate the screw speed at which this type of operation commences.

The theory and development of adiabatic extrusion was considered as early as 1953 when McKelvey presented a paper on the

'Theory of Adiabatic Extruder Operation' at an A.C.S. meeting in Chicago. This work consisted of an analysis of isothermal melt flow supported by experimental results using viscous liquids, and presented a series of equations for adiabatic extruder operations. These were only true for melt extrusion and experimental work was published[51] to extend this theory to normal plasticising extrusion. Three different screw designs were tried and results recorded. These were compared with the theoretical calculations and provide some indication of screw design for adiabatic operation.

In 1955 Gaspar[44] suggested that the speed of a screw of conventional design is severely limited by torsional stress when the extruder is operated without external heaters. Maddock[52] also showed that peripheral speed, drive power and temperature are related in adiabatic operation. His formula $N \sim Q\,\eta\,\gamma$, where $N = l$ power, Q = quantity of material, η = viscosity, and γ = shear gradient, can be simplified to $N \sim n^{1.4}$ where n is the screw speed assuming that the viscosity of polyethylene changes with the 0.6th power of the shear gradient, $\eta \sim \gamma^{-0.6}$. The temperature increase and screw speed relationships are influenced by the screw design, i.e. its compression ratio and corresponding shear gradient and also by the pressure at the die.

Higher screw speeds for adiabatic operation were the subject of a paper by Beck[20,46] and also by Kennaway[53], who operated a 12.7mm (0.5 in) screw at a peripheral speed of 4.1 m/s (160 in/s). At these high speeds the torque on the screw shaft may exceed the strength of the screw where conventional designs are employed. To counteract this the screw must be designed with little or no compression ratio, the feed should be controlled and also the die pressure. With such a screw, i.e. a compression ratio of 1 : 1.5, Beck reports that about 85 per cent of screw length is used for solids conveying and the remainder as a transition and metering zone. This screw worked successfully with polyethylene at peripheral speeds of up to 7.0 m/s (275 in/s). Further experiments with a screw of 35 mm and L/D ratio of 10 : 1, compression ratio 1 : 1, confirm earlier work and indicate higher outputs for polyethylenes of various melt viscosities. Speeds of up to 1500 rev/min were used.

For establishing operating conditions an energy balance equation has to be calculated. McKelvey[54] suggests an energy balance equation

$$Q_p C_p\, dt = \epsilon N^2 \eta\, dl \qquad (3.45)$$

where

$$\epsilon = \frac{\pi^2 D^2}{\tan \phi}\left[\frac{(t - ne)(1 + 3\sin^2 \phi)}{h} + \frac{ne}{\delta}\right]$$

From McKelvey's calculations it is possible to suggest formulae for adiabatic operation

output $$Q = \frac{AN^2l}{R-1} \quad \text{where} \quad A = \frac{\eta_i b \epsilon}{dC_p}$$

pressure $$\Delta_P = \frac{Q\eta_i}{kR}$$

temperature increase $$\Delta t = \frac{1}{b} \ln R$$

power $$Z = dC_p Q \, \Delta t + Q \, \Delta P$$

when η_d is the viscosity at the die, η_i is the inlet viscosity, $R = \eta_i/\eta_d$, K is the die constant, b is the viscosity constant, d is the density, and C_p is the specific heat at constant pressure.

Practical and theoretical results bear a sufficiently close relationship to confirm the validity of the arguments, as proposed[18]. In these experiments screw peripheral speeds of up to 2.5 rev/s were carried out and commenting on this the author suggests that higher screw speeds could be used. Development of the theory and supporting experimental work have also been published[55].

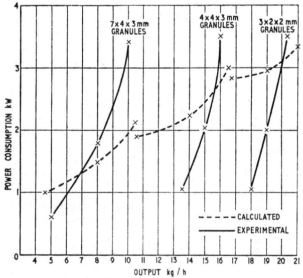

Figure 3.20. Diagram showing the effect of granule size on the output and power consumption in the adiabatic extrusion of nylon

Adiabatic extrusion has been investigated to a limited extent in Russia where Mikhailov *et al.* have reported[56] on an extruder operating at speeds of between 680 and 3570 rev/min processing nylon. A vented screw was employed in which the channel depth of the first metering zone was less than the channel depth of the second metering zone so that there would be no back-flow out of the vent. One of the important features in the extruder screw design was the land width and clearance between it and the barrel; a land width of between 0.04 to 0.06 mm and a clearance of $\delta = 0.1h\sqrt(e)$ mm gave good results both in output and quality. Another important point was the granule size; the larger the granule the smaller the output (*see Figure 3.20*).

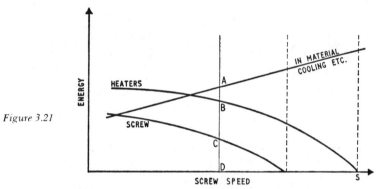

Figure 3.21

The majority of thermoplastics materials can be worked adiabatically on single-screw machines and polyethylene, in particular, can be extruded in this manner at fairly normal screw speeds on correctly designed machines. A considerable amount of experimental work has also been carried out on plastics extruders to investigate the relationship between the energy contributed by the screw and by the heaters, in order to determine the screw speed and conditions at which the system functions without the aid of external heating. The diagram in *Figure 3.21* shows the relationship between the various energy inputs and outputs for a plastics extruder. The line AD shows the conditions prevailing for a certain material temperature and for a given screw speed. The length AD shows the energy leaving the system in the material and in the screw cooling water; and the length AC shows the amount put into the system by the screw. The contribution by external heating is the length BD, and the overlap BC is the energy lost by radiation and conduction. As the speed of the screw increases the line AD moves right until, at a certain speed, the screw is able to supply all the energy needs of the system and this speed is shown at the point S. This type of diagram shows how a normal extrusion system moves towards adiabatic operation as the screw speed increases and indicates that this type of operation could always be achieved if sufficient power were available for the screw.

The power equation 3.42 in Section 3.5 shows that the power input by the screw for a given speed increases as the depth of flight is reduced. It should, therefore, be possible to operate adiabatically at lower screw speeds with a shallow-flighted screw than with a deep one; thus a high compression ratio screw would be specified for that type of operation.

Figure 3.22. An adiabatic extruder in the production of small-diameter polyolefin (Courtesy Alpine A.G., Augsberg)

The interest in adiabatic operation is not merely confined to an academic study of energy relationships; there are certain practical advantages claimed for this method of extrusion. It has been claimed that the thermal efficiency of the extruder is higher and that the material passing through the machine is heated and plasticised more homogeneously when the extruder functions adiabatically and in particular with high screw speed[46].

Figure 3.23 Coextrusion equipment for tubular film using two adiabatic extruders. (Courtesy Paul Kiefel G.m.b.H., Worms)

In order to introduce heat to the plastics material in an extruder by means of conduction from external heaters, it is necessary to maintain a temperature gradient through the material, and this obviously means that the plastics stock is not all at the same temperature. It is, therefore, possible that the output from the machine could have wide temperature variations throughout its mass and that the quality of the product would suffer as a result. On the other hand, if all the heat were supplied to the material by means of shear in the screw flight heat would be generated where required in the material and there would be no need for a temperature gradient. Furthermore, since the temperature gradient through the barrel wall would be in the reverse direction for adiabatic operation, the amount of heat dissipated from the external surfaces would be less and, consequently, a higher thermal efficiency would result.

As mentioned before, polyethylene can easily be worked adiabatically and PVC materials usually generate sufficient heat when passing through an extruder operating at normal speeds to obviate the necessity for external heaters. There are obvious advantages to working PVC materials in that way since their tendency to degrade if a certain critical temperature is exceeded is well known, and any undue temperature gradient maintained to aid conduction would always be liable to bring about this degradation. Screws should, therefore, be designed not only on the basis of output considerations but also from an energy point of view and, if possible, a speed range and output range should be specified upon which to base the initial screw design. In the field of practical application adiabatic extrusion systems have been used with considerable success in the production of small diameter tubes and pipes in a variety of materials, small intricate profiles and particularly in the production of polyethylene tubular film and tubular film composites by coextrusion. In the latter application the small physical dimensions of adiabatically operating extruders enable a number of such machines to be easily attached to a common die system *(Figures 3.22 and 3.23)*.

3.9   Two-stage screw

Two-stage screws are used in conjunction with vented barrels to improve the quality of the extrudate by allowing gases, vapours, etc. to escape from the melt. By this means the predrying of materials can be eliminated and frequently the output can be increased without reduction of quality.

The screw itself has an overall length of approximately 24D or over and consists of two screws in series on the same shaft. The first stage is usually a standard three-zone constanct pitch increasing core diameter screw of about 16D in length, or alternatively a short compression zone type may be used. The metering zone of this stage is usually designed with a normal channel depth about 4D long but smear zones and other devices have also been used to assist degassing. At the end of the first metering zone the core diameter is suddenly reduced for some 4D, depending on the material and difficulty of volatile extraction, during which period of decompression the material tends to froth and all volatiles, gases, etc. escape from the melt and may be extracted by means of vacuum through a vent hole in the barrel or in the screw. The vent zone channel depth is in the ratio of 2.5-4 : 1 to that of the metering zone. Following the vent or decompression zone the material is again compressed and enters the second stage, which consists of a short metering zone with the channel depth greater than that of the first metering zone. This deepening of the channel is necessary to give the second stage, at zero die pressure, a greater pumping capacity than that of the first stage and thus to reduce the risk of exudation of the melt through the vent hole. This factor causes a limitation to be imposed on the die pressure as can be seen from *Figure 3.24*, whereby any die pressure over 17 Mn/m^2 (2500 lbf/in^2) or thereabouts reduces the second-stage output rate to below that of the first stage, which is operating at zero die pressure, and will thus result in back flow out of the vent.

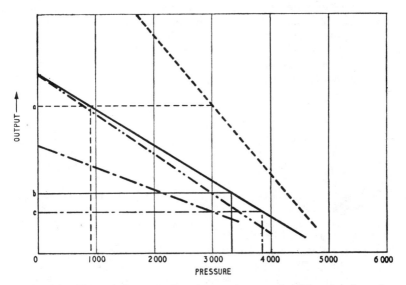

Figure 3.24. Theoretical presentation of output/pressure for different designs of two-stage screws. h_1 = *channel depth on first stage screw;* h_2 = *channel depth on second stage screw.* (a) *Second-stage screw:* x— — — —, h_1 < h_2; y ——.——, h_1 = h_2; z ——.—— , h_1 > h_2; (b) *first-stage screw,*——————. *Example: Die pressure 20 Mn/m^2 (3000 lbf/in^2) with second-stage (x) output=a and first-stage screw pressure 6Mn/m^2 (900 lbf/in^2) minimum; second stage (y) output=b with first-stage screw pressure=23Mn/m^2 (3300 lbf/in^2) minimum; second-stage (z) output=c and first-stage screw pressure=26 Mn/m^2 (3800 lbf/in^2) minimum*

If it is necessary to operate at increased die pressures then a corresponding restriction must be placed on the output of the first stage. This can be done by raising the discharge pressure of stage 1 by means of a valve, in which case it is no longer essential for the second-stage channel depth to be greater than that of the first, and it can be equal or indeed less. Such a system has been described by Grant[57]. Here the first stage ends in a reverse flight which acts as a seal, the melt being bypassed to the deep channel stage through a valve before being taken into the metering section of the second stage.

Instead of having a short compression screw as a second stage, an increasing-core-diameter type of screw can be employed. This calls for greater length and it is therefore necessary to shorten the first stage and to move the venting position towards the feed hopper. The metering section of the first stage can however be much shorter with this type of screw.

With two-stage screws it is in the design of the first stage where trouble can arise in that this stage operates at zero back-pressure and an accurate balance between the feed and metering zones is important. The output of this first stage is controlled by its metering section and the design of the second stage must be based on the overall output figure.

The excellent results reported for decompression screws suggest that this may be a repetition of what is already taking place in injection moulding—namely a distinct separation of the two functions of plasticising or melting, and pressurising.

The operation of this type of screw with the vent hole plugged has not been studied although such a screw has been used successfully for the adiabatic extrusion of nylon. It would seem that it can offer improved quality and homogenisation and may be particularly useful in powder processing. The output rates should be higher because the increased residence time of the melt combined with increased shear will enable the required quality to be obtained at lower die pressures.

An extruder screw of a different design has been proposed by Dulmage[58]. This employs a two-stage screw with a helically grooved milling section between them. A large vent opening in the barrel and a small circumvented channel provide decompression whilst a further set of milling screws are fitted to the discharge end. This screw is primarily designed for mixing and it appears to give very goods results because of the large amount of shear developed in it.

Amongst the many other methods described for venting is the Du Pont system in which the gases and volatiles escape through the hollow core of the screw and not through the barrel[59]. A similar system has been proposed by N.R.M. but with a different screw design. The Danielson system takes the whole idea a stage further and uses two screws and barrels operating in tandem. A later and logical development from the two-stage screw with barrel venting is the splitting of the two stages with separately driven screws to perform the functions of each individual stage with a venting means between the two. Such a system has been used by Barmag in Germany using screws mounted one above the other, by Holt[61] in the United States and others. A more detailed survey of earlier work on the use of two-stage screws has been published by Schenkel[60,70].

3.10 Pressure control

As the theory of extrusion and the behaviour of the material in an extruder became better known so it became apparent that the pressure inside the barrel at the die had a definite influence on quality and quantity of output. This is described in Section 2.7 and specific references to this effect have been made earlier in this chapter.

The back-pressure developed by or at the die is a linear function of the output, assuming a constant speed, so that as a general rule the greater the restriction, the higher the pressure and the less the output. On large nozzles or open dies with low restriction, the expected higher output may need to be curtailed due to non-homogenisation of the melt and poor surface finish resulting from the low heat-generating capacity of low-pressure working and the short residence time.

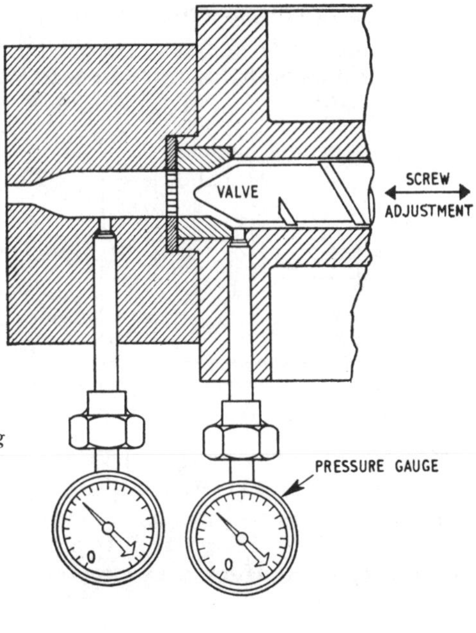

Figure 3.25. Pressure control by means of a conical seating

By restricting the nozzle by means of a valve, improved quality, with a relatively small reduction in output, can be obtained. This is particularly true with small-diameter screws.

Increasing the pressure at the die has the effect of moving the transition zone of the screw further back along its axis and increasing the mixing characteristic and stability of the system. Breaker plate, screen pack and die design all affect the pressure but they cannot provide sufficient control to obtain optimum quality and output conditions. The incorporation of a suitable valve at the outlet is a very effective and convenient way of doing this and at least two methods can be used: (a) axial movement of the screw, (b) adjusting the adaptor channel, sometimes referred to as dynamic and static valving respectively.

Axial movement of the screw either by mechanical or hydraulic means has been incorporated in a number of extruders (*Figure 3.25*). This movement closes or opens the gap between the conical end of the screw and a matching seating. Alternatively the conical seating may be moved in relation to the screw termination. The 'valved' adaptor is adjusted by mechanical means to provide a set opening and consequent pressure. The design of such an adaptor valve must offer controlled restriction to the flow, without creating dead areas. The ordinary bolt type of the valve which is commonly used is not perfect in this respect, but is simple to apply and to control, and is probably quite good enough for all materials except unplasticised PVC (*Figure 3.26*). For accurate control of die pressure whatever system is used it is desirable to include pressure indicators both downstream of the valve and upstream of the breaker plate.

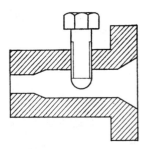

Figure 3.26. Bolt-type valve

Improvement in quality of PVC dry blend extrusion is reported by Sackett and Hankey[62], who also suggest die pressure control on vented extruders as a means of simplifying screw design. Other designs of valved pressure control have been reported[63],[64]. Furthermore Clegg and Huck[65] have reported that by using a valve to increase die pressure blown polyethylene film with improved optical properties was obtained.

Valving by axial movement of the screw in which the conical tip of the screw engages with a ring seating in the end of the barrel has limitations in that when high pressures are required positive contact between the screw and seating could occur with consequent damage. An alternative design (*Figure 3.27*) to overcome this has been developed and tested by Sponaugle[66] and the results indicate that this construction gives higher outputs, and better mixing properties than the ordinary screw-down bolt type valves. This is due to the extra working and shear which the material receives in the valve and its increase in temperature. The depth of rings recommended is 38 mm (1.5 in) on an 85 mm (3.5 in) 20 : 1 L/D screw as at a lesser depth the heating, pressure and efficiency decrease.

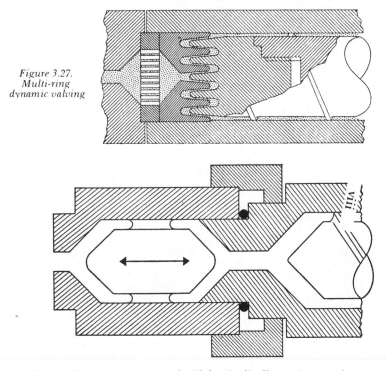

Figure 3.27.
Multi-ring
dynamic valving

Figure 3.28. Die pressure control with longitudinally moving torpedo

Kreft and Doboczky[67] suggest several methods for controlling die pressure including an axially moving torpedo in the die

(*Figure 3.28*). They also report the relationship found by experiment between pressure, screw speed and output. As an alternative to a valve for controlling the pressure, however, Maddock[68] has suggested that a fixed orifice in the form of a breaker plate or an inserted adaptor plate could perform the same function and it is apparently not too difficult to design such an orifice to give a prescribed pressure.

The importance of pressure control is well brought out in a paper by J. van Leeuwen, in which details are given of an experiment carried out at Delft, where the extruder was fitted with a large number of transducers in the barrel and die to give melt pressure records with various materials and screw designs[71]

3.11   Intermittent extrusion

Controlling the flow of material at the die by means of a valve is often an important part of the successful use of an extruder for the production of parisons for bottle blowing. The design of such valves does not present any serious difficulty, the need for free flow being the principle criterion. Plug and needle valves have both been used successfully and their design is discussed by Bernhardt[69], who also considers the effect of their use on the material in the barrel.

REFERENCES

1   ROWELL, H.S. and FINLAYSON, D., 'Screw Viscosity Pumps', *Engineering. Lond.*, 114, 606 (1922); 126, 249, 385 (1928)
2   REYNOLDS, O., *Phil. Trans. R. Soc.*, 177, 157 (1886)
3   NAVIER, C.L.M.H., *Mem. Acad. Sci. Inst. Fr.*, 6 (1822)
4   POISSEUILLE, J.L.M., *C. r. hebd. Seanc. Acad. Sci., Paris*, 15, 1167 (1842)
5   ROGOWSKY, Z., 'Mechanical Principles of the Screw Extrusion Machine', *Engineering. Lond.*, 162, 358 (1946)
6   PIGOTT, W.T., 'Pressures Developed by Viscous Materials in the Screw Extrusion Machine', *Trans. Am. Soc. mech. Engrs*, 73, 947 (1951)
7   GRANT, D. and WALKER, W., *Plast..Prog., Lond.*, 245 (1951)
8   EIRICH, F.R., *Proc. Instn mech. Engrs*, 156, 62 (1947)
9   MAILLEFER, C., 'Etude theoretique et experimentale sur le fonctionement des boudineuses', *These de Doctrat*, Lausanne (1952); abridged form 'Analytical Study of the Single Screw Extruder', *Br. Plast.*, 27, 394 (1954)
10   GORE, W.L. *et al.*, 'Principles of Plastic Screw Extrusion', *Ind. Engng Chem.*, 45, 969 (1953); *S.P.E. J1*, 9, 3 (1953)
11   MESKAT, W., 'Theory of the Flow of Materials in Worm Machines (Pt I)', *Kunststoffe*, 45, 87 (1955)
12   SPENCER, R.S. and DILLON, R.E., 'The Viscous Flow of Molten Polystyrene', *J. Colloid Sci.*, 3, 163 (1948)

13 SQUIRES, P.H., PATON, J.B., DARNELL, W.H., CASH, F.M. and CAR-
    LEY, J.F., in *Processing of Thermoplastics*, Ed. Bernhardt, E.C., Reinhold,
    New York, for the Society of Plastics Engineers (1959)
14 SCHENKEL, G., *Schneckenpressen fur Kunststoffe*, Carl Hanser Verlag,
    Munich (1959)
15 DARNELL, W.H. and MOL, E.A.J., 'Solids Conveying in Extruders', *S.P.E.
    Jl*, **12**, 20 (1956)
16 RAKHMANOV, V.S., 'Determining the Output of Extruders', *Soviet Plast.*,
    5, 42 (1961)
17 SCHIEDRUM, H.O. and DOMININGHAUS, H., 'The Performance of
    Single Screw Extruders in the Processing of Polyolefins and PVC', *Plastics,
    Lond.*, **26**, 81 (1961)
18 JACOBI, H.R., 'Flow Characteristics in the Single Screw Extruder (Pt II)',
    Int. Plast. Engng, **2**, 264 (1962)
19 STREET, L.F., 'Plastifying Extrusion', *Int. Plast. Engng*, 1, 289 (1961)
20 BECK, E., 'Betriebsverhalten und Praxisergebuisse von schnellen fenden
    Schneckenpressen', *Kunststoffe*, **52**, 213 (1962); 'High Speed Extruder Charac-
    teristics', *Int. Plast. Engng*, **2**, 316 (1962)
21 UNDERWOOD, W.M., 'Experimental Methods for Designing Extrusion
    Screws', *Chem. Engng Prog.*, **58**, 59 (1962)
22 KLEIN, I. and TADMOR, Z., 'Computer Design of Plasticating Extruder
    Screws', *Mod. Plast.*, **46**, 166 (1969)
23 STEVENS, M.J., 'The Melting Process in Screw Extruders', *Plast. Polym.*, **38**,
    107 (1970)
24 DONOVAN, R.C., 'Theoretical Melting Model for Plasticating Extruders',
    Polym. Engng Sci., **11**, 247 (1971)
25 SQUIRES, P.H., 'Screw Extruder Pumping Efficiency', *S.P.E. Jl*, **14**, 24
    (1958)
26 MADDOCK, B.H., 'Effect of Wear on the Delivery Capacity of Extruder
    Screws', *S.P.E. Jl*, **13**, 443 (1957)
27 BARR, R.A. and CHUNG, C.I., 'Effects of Radial Screw Clearance on
    Extruder Performance', *S.P.E. Jl*, **22**, 71 (1966)
28 MOHR, W.D., SAXTON, R.L. and JEPSON, C.H., 'Mixing in Laminar
    Flow Systems', *Ind. Engng Chem.*, **49**, 1855 (1957)
29 STRUB, R.A., 'The Theory of Screw Extruders', *Proc. Second Midwestern
    Conf. of Fluid Mechanics, Ohio State University, 1952*, p. 48
30 MORI, Y., OTATAKE, N. and IGARASHI, H., 'Screw Extrusion Process for
    Forming Plastic Materials', *Chem. Engng, Tokyo*, **18**, 22 (1954)
31 MADDOCK, B.H. and SMITH, D.J., 'Extruder Design by Computer Prin-
    tout', *S.P.E. Jl*, **28**, 12 (1972)
32 RYDER, L.B., 'Some Comments on Vented Extruder Design', *S.P.E. Jl*, **17**,
    731 (1961)
33 MARTIN, B., 'Theoretical Aspects of Screw Extruder Design', *Plast. Polym.*,
    38, 113 (1970)
34 TOMIS, E., 'Improvements in Extruding PVC on Screw Presses', Gottwaldov
    Research Institute, 1953
35 *idem ibid*
36 MADDOCK, B.H., 'Pressure Development in Extruder Screws', *S.P.E. tech.
    Pap.*, **6** (1960); *S.P.E. Jl*, **16**, 373 (1960)
37 MADDOCK, B.H., 'An Improved Mixing-Screw Design', *S.P.E. Jl*, **23**, 23
    (1967)
38 PFLUGER, R., 'Polyamide in der extruder Verarbeitung', *Kunststoffe*, **52**,
    273 (1963)

39 TOLL, K.G., 'Shape Extrusion Studies of Zytel Nylon Resin', *S.P.E. tech. Pap.*, 3 (1957); *S.P.E. Jl*, 13, 17 (1957)

40 GARBER, J.F. and CASSIDY, R.T., 'Performance Characteristics of Metering Type Screws on Acrylic Materials', *S.P.E. tech. Pap.*, 8 (1968)

41 KREFT, L. and DOBOCZKY, Z., 'Steigerung der Forderleistung von Schneckenpressen', *Plastverarbeiter*, 13, 348 (1962)

42 MOHR, W.D., MALLOUK, R.S. and BOOY, M.L., private communications to Squires, P.H., cited in BERNHARDT, E.C. (Ed.), *Processing of Thermoplastic Materials*, Reinhold, New York, for the Society of Plastics Engineers (1959)

43 GORE, W.L. and MCKELVEY, J.M., 'Theory of Screw Extruders', in *Rheology — Theory and Applications*, Vol. 3, Academic Press, New York (1958)

44 GASPAR, E., *Plast. Prog., Lond.*, 151 (1955)

45 idem ibid

46 BECK, E., 'Improvements of Screw Extruder Efficiency by Increasing Peripheral Speed', *Kunststoffe*, 49, 315 (1959)

47 CARLEY, J.F. and MCKELVEY, J.M., 'Extruder Scale-up Theory and Experiments', *Ind. Engng Chem.*, 45, 989 (1953)

48 CARLEY, J.F., 'Single Screw Pumps for Polymer Melts', *Chem. Engng Prog.*, 58, 53 (1962)

49 KIRBY, R., 'Process Dynamics of Screw Extrusion', *S.P.E. tech. Pap.*, 8 (1962)

50 KRUEGER, W., 'Experimental Illustration of Dynamic Extrusion Theory', *S.P.E. tech. Pap.*, 8 (1962)

51 MCKELVEY, J.M. and BERNHARDT, E.C., 'Adiabatic Extrusion of Polyethylene', *S.P.E. Jl*, 10, 22 (1954)

52 MADDOCK, B.H., 'Power and Heat Energy Relations in Polyethylene Extrusion', *Kabelitems*, No. 105 (1958)

53 KENNAWAY, A., *Plast. Prog., Lond.*, 149 (1957)

54 MCKELVEY, J.M., 'Analysis of Adiabatic Plastics Extrusion', *Ind. Engng Chem.*, 46, 660 (1954)

55 GLYDE, B.S. and HOLMES-WALKER, W.A., 'Screw Extrusion of Thermoplastics', *Int. Plast. Engng*, 2, 596 (1962)

56 MIKHAILOV, P.A., MAYLSHEV, P.N. and DUPLENKO, Y.U., 'High Speed Extruder for Processing Polyamides', *Soviet Plast.*, 1, 42 (1961)

57 GRANT, D., 'Developments in Extrusion Machinery: Valved and Vented Extrusion', *Trans. Plast. Inst.*, 29, 130 (1961)

58 U.S. Pat 2 753 595 (24 July 1953)

59 BERNHARDT, E.C., 'The Vacuum Extruder Screw', *S.P.E. Jl*, 12, 40 (1956)

60 SCHENKEL, G., 'Vented Single Screw Extruders', *Int. Plast. Engng*, 2, 384 (1962)

61 HOLT, J.E., U.S. Pat. 2 836 851 (9 September 1955)

62 SACKETT, R.D. and HANKEY, E.H., 'Important Advances in Extrusion Technology', *S.P.E. Jl*, 13, 49 (1957)

63 MADDOCK, B.H. *et al.*, 'Controlled Pressure (Valve) Extrusion', *Rubb. Plast. Age*, 39, 1075 (1958); Annual Signal Corps Wire and Cable Symposium, Asbury Park, U.S.A. (Dec. 1957)

64 SMART, G., 'Recent Advances in Extruder Design for Operation at High Melt Temperatures', *Int. Plast. Engng*, 2, 108 (1962)

65 CLEGG, P.L. and HUCK, N.D., 'The Effect of Extrusion Variables on the Fundamental Properties of Tubular Polyethylene Film', *Plastics, Lond.*, 26, 114 (1961)

66 SPONAUGLE, H.E., 'Dynamic Valving versus Static Valving, *S.P.E. tech. Pap.*, 8 (1962)

67   KREFT, L. and DOBOCZKY, Z., 'Erfahrungen uber die Produktivitatssteiger-
ung durch bruckreguliertes Strangpressen', *Plastverarbeiter*, **13**, 127 (1962)
68   MADDOCK, B.H., 'Fixed Orifice Pressure Control for Extruders', *Plast.
Technol.*, **6**, 41 (1960)
69   BERNHARDT, E.C., 'Valved Extrusion', *S.P.E. tech. Pap.*, **3** (1957)
70   SCHENKEL, G., *Schneckenpressen fur Kunststoffe*, Carl Hanser Verlag,
Munich (1959), p. 208
71   VAN LEEUWEN, J., 'Fundamental Measurements and Control in the Extrusion Process', *Plast. Polym.*, **42**, 104 (1974)

4

Multi-screw Extrusion Machines

4.1 General and historical

As explained in Chapter 2 and elsewhere, thermoplastics materials were first extruded on machines which had been designed and developed for the rubber industry. By a process of evolution, the single-screw extruder has been adapted over the years into a machine suitable for thermoplastics and which, although vastly different in detail, is basically similar to the early rubber forcers.

At the time when the thermoplastics materials were becoming more freely available, i.e. in the middle to late 1930s, a school of thought arose, chiefly in Continental European countries, which suggested that an entirely new type of extruder was required to handle these materials properly[1]. The supporters of this view felt that the properties of thermoplastics were so far removed from those of rubber that the basic principles of the rubber machine could never be suitably adapted. The new materials required high pressures, and showed a greater tendency to 'slip' or stick to the screw and thus to rotate with it. Furthermore, they needed to be worked at much higher temperatures than was required with rubber, and also tended to degrade very quickly at these temperatures. This latter difficulty was, of course, particularly in evidence under conditions of slip.

This school of thought suggested therefore that a more positive pump mechanism was required which led to the adoption of the intermeshing-thread twin- or multi-screw principle.

The early history of multi-screw extruders is very obscure but it is believed that this system was experimented with and used for the processing of ceramics, soap, waxes, foodstuffs and other materials of a like nature during the latter half of the nineteenth century. Certainly it is on record[2] that the Manchester firm of engineers Follows and Bate Ltd supplied double-screw 'mincing machines' for making sausage meats as far back as 1880, and a U.S.A. patent application on a twin screw in the name of F.F. Pease was registered[3] in 1933, and assigned to Lever Bros.

The use of the multi-screw principle for thermoplastics, however, is believed to have originated in Italy where the background of experience acquired in the manufacture of spaghetti, macaroni, etc. was undoubtedly a contributory factor. In the middle or late 1930s Roberto Colombo of Turin produced a successful two-screw arrangement, in which the screws intermeshed and thus formed a positive pump. This first machine consisted of screws of the same length, with identical threads and constant pitch, rotating in opposite directions. The arrangement ensured a positive feed and, by virtue of the mutual wiping action of the thread flanks, prevented the material from adhering to the screws. It also became possible to achieve high pressures at the die and to maintain these pressures consistently.

At an early date Colombo abandoned the principle of opposite rotating screws and adopted the patented similarly rotating system as used in modern machines of this type[4] which are manufactured in England by R.H. Windsor Ltd[5] (*Figure 4.1*).

Figure 4.1. Modern twin-screw extruder employing screws rotating in the same direction. (Courtesy R.H. Windsor Ltd)

Another name which occurs early in the history of the multi-screw extruder is that of Carlo Pasquetti, whose two-screw machines have also been developed into a modern version along parallel lines[6] the principle of opposite rotating screws being in this case retained (*Figure 4.2*).

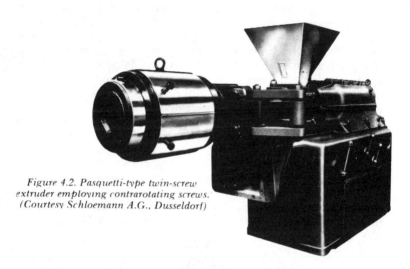

Figure 4.2. Pasquetti-type twin-screw extruder employing contrarotating screws. (Courtesy Schloemann A.G., Dusseldorf)

Apart from the Windsor (Colombo) and Pasquetti machines a large number of other multi-screw extruders based usually on the contrarotating screw system but each having their own special design features have been developed although not all of them achieved commercial success. Some of the more interesting of these machines are or were the Anger; the AGM; the Mapre; the Kestermann; the Trudex; the Leistritz; the Darex and the Jumex (four screws)[7]. Although all of the multi-screw machines mentioned so far are or were recommended for both compounding and direct extrusion several makers have concentrated on compounding and have used the twin screw principle in machines which were designed specifically for this purpose. Perhaps the best known of these are manufactured by Krauss-Maffei and Werner-Pfleiderer in Germany whilst in the U.S.A. Welding Engineers Inc. of Norristown produce a range of two-screw compounding machines in which the screws do not intermesh[8]. Certain of these machines are discussed in greater detail in the following sections.

As with single-screw machines, it was found necessary to introduce a compression system or transition zone in the multi-screw arrangements to allow for the change in bulk factor of the feed material as it passes from the hopper to the die. On the Colombo-type machine, this compression ratio is brought about by changing the pitch and diameter of the screws in three distinct zones. On other machines the compression is achieved by changing the clearances between the flanks of the intermeshing threads, or by changing the flight pitch in zones and a gradual taper system is also used.

A further problem which beset the multi-screw machine designer in the early days was that of trapped air or gases which were unable to escape back to the hopper because of the positive intermeshing of the screw flights. This difficulty was approached in different ways by the various manufacturers: by carefully adjusting the clearances between the screw lands and the barrel wall or between the flanks of the threads, and even by cutting connecting gashes across the threads.

It also became necessary in some cases to fit special feeding devices to these machines so that the material could be metered into the feed section of the screws in order to eliminate any tendency to overpack the threads.

Finally, in order to increase the output of the machines the screws have been made larger in diameter and also the number of screws working together has sometimes been increased. This has resulted in a three-screw arrangement in which the centres of the three equal-diameter screws are located at the apices of an equilateral triangle, and a four-screw machine.

Experimental set-ups have also been made in which five or more screws were arranged to intermesh in a plane, or in a circle, for the extrusion of special products. With the screws arranged in a plane and to feed a common die, the setup was thought to be ideal for the extrusion of wide sheet in difficult materials whilst the circular arrangement was obviously intended for the production of large-diameter pipes. Similar arrangements but in which the screws do not intermesh have also been noted[9].

4.2 Theoretical considerations

There is no body of theoretical work in existence dealing with the mechanism of the multi-screw machine which is in any way comparable with that for the single screw. This dearth of information is mainly due to the extremely complex process which takes place inside a multi-screw machine. No convenient approximation comparable to the channel flow equations for the single-screw machine can be applied in this case and attempts so far to describe the processes involved have largely been limited to verbal pictures.

Material which is fed to the screws of a two-screw extruder is immediately established in discrete C-shaped sections about the screws, as can be seen in *Figure 4.3*.

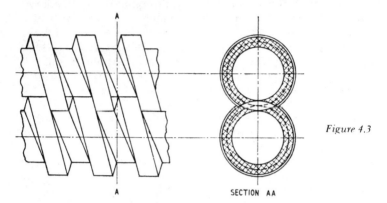

SECTION AA

Figure 4.3

Because of the intermeshing of the two screws the material cannot be easily transferred from one of these sections to another, and it is inevitably carried forward through the barrel. These discontinuous sections, therefore, prevent the use of relationships based on continuous shear, and the material trajectory at the ends of these sections is so complex as to prevent the establishment of a simple theory.

The screw speeds in multi-screw machines are normally much lower than those in single-screw machines. This would appear to mean that much less energy is consumed in the screw flights, and that the energy input to the material is effected largely by conduction from external heaters.

The extruder, then, can be said to operate as a positive pump, and this fact at least enables a relationship to be established for the output of the machine. Because of the discontinuous state of the material being conveyed there cannot be a term comparable in any way to the pressure back-flow in single-screw output equations. The output is, therefore, directly proportional to the speed of the screws and also to the volume of each discrete packet.

A delivery equation may be written

$$Q = kNM_s \qquad (4.1)$$

where Q is the volumetric discharge, k is a constant, N is the speed of the screws, and M_s = volume of a discrete section of screw flight.

Since two sections of material are delivered for every revolution of the screws, $k = 2$.

Furthermore, M_s can be written approximately as

$$\frac{\pi}{2}ht(D - h)q \tag{4.2}$$

where h is the depth of flight, t the pitch, and D the external diameter of the thread. The factor q is always less than unity and expresses the fraction of the screw-thread pitch line unobscured by meshing. In *Figure 4.3* the unobscured pitch line is shown dotted and if this length is c then

$$\frac{c}{\pi(D - h)} = q$$

Therefore, substituting equation 4.2 in equation 4.1, the equation for the output becomes

$$Q = \pi q N h t(D - h) \tag{4.3}$$

This equation can only apply to the die end of the screws, where the material completely fills the screw flights and, since the material is able to flow, allowance must be made for the leakage flow across the top of the screw lands. The term for this leakage flow is similar to that obtained for the same flow in the single-screw machine. The leakage is thought of as a pressure flow through a wide, thin slit, which can be expressed

$$Q = \frac{wh^3}{12\eta}\left(\frac{\mathrm{d}p}{\mathrm{d}z}\right) \tag{4.4}$$

where Q is the volumetric flow, w is the width of the slit, h is the depth of the slit, and $\mathrm{d}p/\mathrm{d}z$ is the pressure gradient across the slit. The width of the slit around the two screws is given approximately by

$$w = 2q\pi D$$

and the depth of the slit is the clearance between the screws and the barrel wall, say δ. Substituting these values in equation 4.4, inserting the eccentricity factor E, and adding this term algebraically to equation 4.2 gives the complete expression for the output

$$Q = \pi N q h t(D - h) - \frac{\pi q D E \delta^3}{6\eta}\left(\frac{\mathrm{d}p}{\mathrm{d}z}\right) \tag{4.5}$$

As mentioned above, this equation refers to the section of the screws in which the material is viscous and completely fills the screw flights. The determination of dp/dz, the pressure profile along the screw, is very difficult in the case of a twin-screw machine since a continuous build-up of pressure cannot strictly be envisaged owing to the discontinuity of the material in the screws. Thus a pressure build-up can only take place in the last few flights of the screws where sufficient leakage across the top of the flights and between the flanks of the thread allows a gradient to establish itself.

Equation 4.5 can be written

$$Q = AN - BP \qquad (4.6)$$

where A and B are constants and P is the total pressure build-up at the end of the screws. The expression 4.6 is a straight-line relationship between Q and P, and the line has a negative slope depending on the magnitude of the leak flow. Because δ is always very small the slope of these screw characteristics will always be small and the lines will be very much flatter than the corresponding characteristics for the single-screw machine shown in *Figure 3.8*. Furthermore, the term B of equation 4.6 is completely independent of the screw profile. Thus the twin-screw machine is only slightly sensitive to die changes, and the fall-away in output resulting from fitting a smaller die to the machine depends only on the magnitude of the leak flow as shown in *Figure 4.4*.

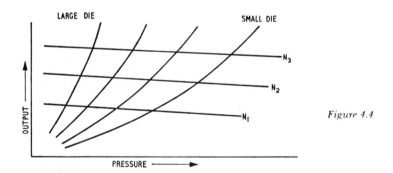

Figure 4.4

The problem of the power requirement for a multi-screw machine cannot yet be solved since little is known about the very complicated shear patterns which are produced by intermeshing screws. The material behaviour of each of the discrete sections can probably be linked to flow in a narrow deep channel at the end of which is a very small restriction formed by the intermeshing of the two screws. Because of the cross section of the channel, no one-dimensional approximation could be used in the differential equation of flow, and the velocity distribution obtained would, therefore, be more complex than that used in the simplified flow and power equation for the single-screw machines. However, assuming that a velocity distribution could be obtained, an expression for the power requirement of the machine could theoretically be derived.

Because of the low speeds of the screws of a multi-screw machine, relatively little energy is introduced to the material in shear, thus most of the heating is done by an external transfer system. The amount of heat which can be introduced into the material by such a system, therefore, is obviously time-dependent, and the material has to remain in the extruder for a certain minimum interval. This fact puts a definite upper limit to the rate of working of a multi-screw machine, or of any other machine which depends mainly on conducted heat.

Some discussion in detail on output, thrust bearing and design, axial force, driving torque moment and coefficient of friction for twin screw extruders has been published by Mäkelt. This deals both with practical and theoretical design considerations as well as output and general constructional features on these machines[10].

4.3 Modern multi-screw machines

The number of variables available on a twin-screw machine is obviously greater than that available on the single-screw machine.

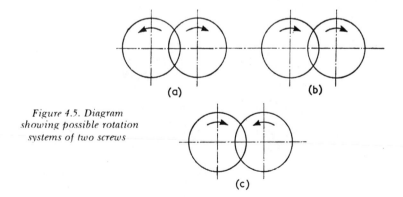

Figure 4.5. Diagram showing possible rotation systems of two screws

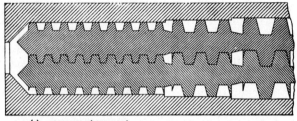

(a) COLUMBO (WINDSOR)

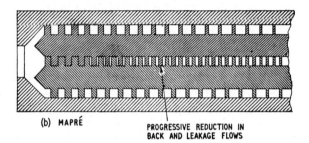

(b) MAPRÉ

PROGRESSIVE REDUCTION IN
BACK AND LEAKAGE FLOWS

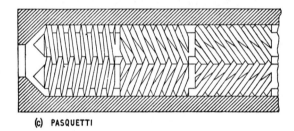

(c) PASQUETTI

ROTATING HELICALLY GASHED DISCS

(d) TRUDEX   DEEPLY INTERMESHED
SCREWS, NO NORMAL
COMPRESSION

Figure 4.6. Illustration showing typical commercial two-screw arrangements

It is not only possible to alter the thread profile, compression ratio and length/diameter ratio, but also to vary the relative rotation of the screws and the clearances between them. There are three possible rotational arrangements as shown in *Figure 4.5*, and there are modern extruders based on each of these systems. The arrangements shown in *Figure 4.5a* is that used in the machines developed by Pasquetti, Trudex, Mapre, and others, whilst *Figure 4.5b* shows the direction of rotation of the screws of the modern Colombo-type machine. Finally, *Figure 4.5c* the third possible system of rotation which is used in the non-intermeshing screw machines of Welding Engineers Inc.

Certain advantages from the point of view of mixing and compounding are claimed by the advocates of the different systems used but it is obvious that an extruder can work satisfactorily whichever method is adopted.

Many different screw configurations have been developed in order to achieve a so-called compression ratio, and each manufacturer appears to favour a different system. In the Colombo, *Figure 4.6a*, the screws are cut in three zones, each of which has a different pitch as well as a different root and outside diameter. The rear section of the screws nearest to the hopper has a coarse pitch and the largest outer and smallest inner diameter. The middle section is slightly longer than the rear section, has a slightly smaller pitch, and the outer and inner diameters are somewhat smaller and larger respectively. The front section of the screw is the longest, has the smallest pitch, and its depth of flight is further decreased.

As can be seen from the diagram, the threads in all the three sections are very deeply cut. The flanks are tapered, and the screws intermesh almost to the full thread depth along their whole length. A further feature is that originally no cooling channels were provided in the screws although modern machines now frequently have this feature. The barrel of the machine has a figure eight bore and is cut in three steps to accommodate the three sections of the screw, and the clearance above the lands is of the order of a tenth of a millimetre (a few thousandths of an inch.)

The Mapre system consists basically of screws which are square cut as in *Figure 4.6b* and the compression is obtained by progressively decreasing the clearances between the flanks of the threads. Mapre have in fact a number of designs of screw for their machines giving different degrees of friction and compression depending on material requirements. In some cases flats are machined locally at a predetermined position along the screw length to allow gases to escape back to the feed orifice and to give additional mixing and mastication of the material. Apart from the screws themselves,

Mapre have also developed a number of interchangeable screw tips whose configuration may be conical, conical with flats, spiral or of other form. The selection of the screw tip most suited to the material characteristics is said to have a marked effect on the operation of these machines.

The arrangement used by Pasquetti, *Figure 4.6c* is similar to that used in the Colombo system, inasmuch as the screws are cut in three sections, with gaps between them. However, the screws of the Pasquetti machine have a parallel outside diameter, and the compression is obtained by decreasing the pitch and number of starts progressively along the three sections. The thread form is similar to an Acme profile, and cooling water is not usually provided on the screws.

On the early Trudex machines no compression ratio was provided and the thread of roughly square form did not vary at all along the screw. Subsequent design changes, however, resulted in the machine being fitted with screws having a certain compression ratio. A series of gashed discs were mounted on the die end of the screws which acted as mixers and compounders, to cut and masticate the softened material[11]. The screws were bored for water or air cooling (*Figure 4.6d*).

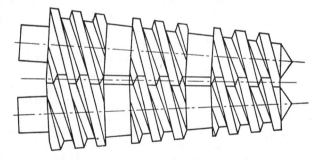

Figure 4.7. Two-screw arrangement with converging axes

A further method of obtaining compression, which is not quite so well known as the methods outlined above, is shown in *Figure 4.7*. In this arrangement a pair of tapered screws are used in which three-zoned threads are cut. The conical shape of the screws and the bore into which they fit enables a certain compression to be obtained in a gradual manner, as opposed to the discontinuous system of the Colombo and Pasquetti screws. In addition to providing a method of obtaining compression, the angled disposition of the centre lines of the screws also enables larger thrust bearings to

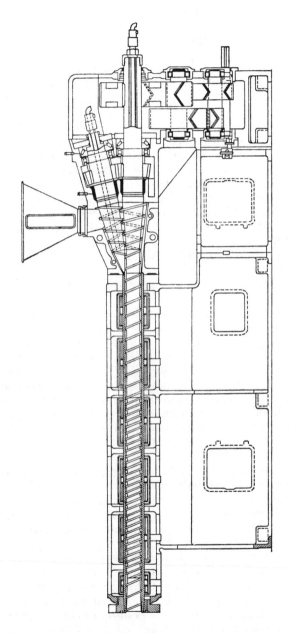

Figure 4.8. Sectional drawing of Berstorff KD120 extruder showing twin-screw feed. (Courtesy Hermann Berstorff, Hanover)

be fitted than would be possible with a parallel system. Since the bearing limitations to twin-screw extruders constitute a major design problem the slight advantage which this arrangement appears to give is a worthwhile consideration. Other designs of twin-screw extruders but employing continuous tapered screws have also been noted[12] and the modern AGM machines, now manufactured by Cincinnati Milcron, Austria G.m.b.H., are based on this concept.

Further interesting developments or extensions of the multi-screw systems of extrusion which should be mentioned are the Eck Mixtruder now manufactured by Krauss-Maffei and incorporating mulling sections on the screws[13], the Berstorff types KD120 which use subsidiary conical feed screws arranged above the main screws (*Figure 4.8*), the Welding Engineers' machine with its screws of different lengths incorporating a reverse flight as the termination of the shorter screw[8], the Werner Pfeiderer with interchangeable kneading discs, the Pasquetti Bihelicoidal compounder[14], and finally the Anger two-stage twin-screw machine (*Figures 4.9* and *4.10*).

The Pasquetti Bihelicoidal compounder consisted of three pairs of meshing double-helical gears placed vertically in a closely fitting oil-heated housing. The material was fed to the top set of gears where it was picked up by the gear teeth and conveyed outwardly around the housing to the second and third pairs successively to emerge finally via a conventional breaker plate/screen pack system at the base of the housing. The pitch of the gear teeth was reduced progressively from the top to the bottom set: this reduction providing the compression ratio of the machine (*Figure 4.11*). Although an interesting and ingenious concept, it is not known just how successfully the Bihelicoidal compounders performed in practice and it is believed that the machines may now be out of production.

The Anger two-stage twin-screw machine mentioned above uses two pairs of intermeshing screws arranged in series with the output from the first pair, known as the 'plasticising' screws, feeding directly into the second pair or 'discharge' screws. Both pairs of screws are separately powered with variable-speed drives and by adjusting the speed ratio it is possible to ensure optimum filling of the discharge section without overloading the plasticising unit. *Figures 4.9* and *4.10* give a schematic view showing the internal construction and a view of the complete machine respectively.

Over the years since this book was first published there have been many detail changes in the design of twin-screw extruders: L/D ratios, thrust bearing systems and screw configurations have

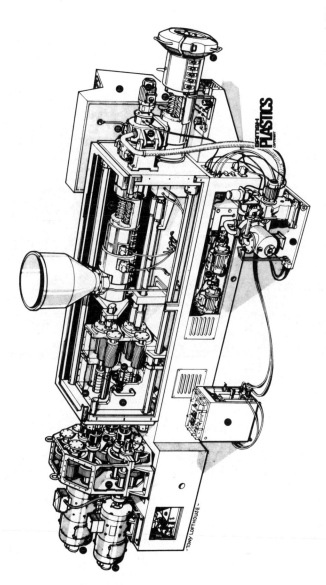

Figure 4.9. Schematic view of the Anger two-stage twin-screw extruder. (Courtesy Cincinnati Milacron. Austria G.m.b.H.)

been improved, for example, and the machines increased in size, robustness and output capacity; however, the principles of operation in general remain the same.

Nevertheless, before leaving this section it will perhaps be interesting to note certain of the other changes which have occurred in the twin-screw extruder manufacturing field during the above period.

Figure 4.10. The Anger two-stage twin-screw extruder. (Courtesy Cincinnati Milacron, Austria G.m.b.H.)

The brothers Wilhelm and Anton Anger, who were responsible in the early 1950s for the development of the highly successful machines of the same name decided some years ago to go their separate ways, the original Anger twin-screw concept and the Anger company being retained by Wilhelm whilst Anton started a separate company and developed the AGM machines, which have already been referred to, and which use intermeshing screws of a conical configuration. The Wilhelm Anger company and the manufacturing rights to the Anger machines were acquired in the mid 1960s, first 50 per cent and finally 100 per cent by the American company Bemis Corporation of Minneapolis and Sheboygan. Finally, in the early 1070s Bemis sold the company and the rights to Cincinnati Milacron.

Cincinnati had also acquired about one year earlier Anton Anger's AGM company and the rights to this machine. With the merger of both Anger Companies in Cincinnati Milacron Austria G.m.b.H Cincinnati Milacron is perhaps now one of the most important manufacturers of twin-screw extruders.

Mapre (Machines et Appareils de Précision S.A.) is now part of the Swiss Studle organisation.

Rolf Kestermann Maschinenfabrik of Bad Oeynhausen, West Germany, the manufacturer of the Kestermann machine, was acquired on 1 April, 1971, by Rheinstahl A.G., whilst the Trudex machines are believed to be no longer in production.

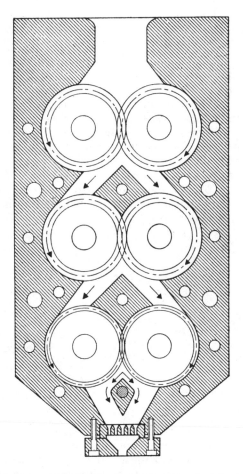

Figure 4.11. Diagrammatic arrangement showing the principle of the Pasquetti bihelicoidal gear extruder.

The Pasquetti machine has also seen changes. In the 1960s the manufacturing and sales rights of these machines, outside Italy, were acquired by the German engineering company Schloemann A.G. of Düsseldorf, who re-engineered the machines and gave them the name 'Bitruder', under which title the machines enjoyed considerable success. In 1971 however the manufacturing and sales rights to the Schloemann Bitruders were acquired by Reifenhauser, one of the largest European manufacturers of single-screw extruders.

4.4 Comparisons between the characteristics of multi-screw and single-screw machines

4.4.1 OUTPUT

The form of the theoretical relationship between the output pressure and the screw speed for multi-screw and single-screw machines was discussed in previous sections. From a comparison of the equations it appears that the multi-screw machine is the least sensitive to die variations, and its output would remain fairly constant for a given screw speed over a wide range of die apertures. From this point of view, therefore, a twin-screw extruder is similar to a single-screw machine with a screw of high compression ratio, or shallow metering section.

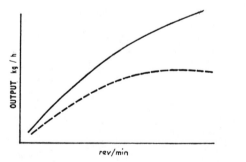

Figure 4.12

The plot of output against screw speed for a multi-screw machine tends to a straight line much more closely than for the single-screw machine. For the multi-screw extruder the volumetric efficiency of the screws is fairly high, and is almost independent of pressure. In the case of the single-screw machine, on the other hand, volumetric efficiencies are always low and are highly pressure-dependent. The reason for this lies in the importance of the pressure flow term in the single-screw output equation and the complete absence of this term from the corresponding equation for the twin screw. A typical output/screw-speed plot for a single-screw machine is shown with an exaggerated convexity in *Figure 4.12* and the deviation from the straight line increases with screw speed because the pressure increases also with screw speed, and thus the pressure back-flow term becomes greater. When very difficult feeding materials such as powders and very heavily externally lubricated granules are used, the forward thrust is so small that the back flow approaches it in magnitude, excessive slip then takes place, and the curve in *Figure 4.12* has a very definite upper limit as shown by the dotted line.

4.4.2 POWER REQUIREMENTS

The equations in Chapter 3 for the power requirements of a single-screw machine showed that the power necessary to turn the screw was proportional to the square of the screw speed. Thus

$$Z = AN^2 \tag{4.7}$$

From the output equation the output is almost directly proportional to the screw speed and in general it is reasonable to write

$$Q = BN \tag{4.8}$$

Substituting 4.8 in 4.7 gives the power requirement for a given output, and

$$Z = \frac{A}{B^2}Q^2 \tag{4.9}$$

From this equation it may be seen that the power is proportional to the square of the extruder delivery and, since energy appears in the material as heat, the extrudate becomes hotter as the screw speed increases, provided sufficient power is available to turn the screw.

The same cannot be said for multi-screw machines. Because of the deep screw flights the power requirements are relatively low and the greater part of the heating is effected by conduction from barrel heaters. The material heating and, therefore, the allowable rate of throughput is thus time-dependent and there exists for all multi-screw machines a definite upper speed limit which depends on the thermal properties of the material being extruded.

The above considerations appear to rule out the possibility of adiabatic operation on multi-screw machines and the advantages obtained for this mode of operation will apparently remain peculiar to single-screw machines. On the other hand it is argued that the conduction method of heat input is the most easily controlled and the most versatile system, which perhaps cannot yet be claimed for the shear method of heat generation.

4.4.3 GENERAL COMPARISONS

From the purely practical point of view the arguments for and against multi-screw extruders as compared with single screw machines are difficult and mildly controversial. On the one hand the makers of multi-screw machines recommend their equipment for mixing and compounding and for direct extrusion from uncompounded powder blends—a recommendation based on well authenticated practical proof—and on the other hand the convinced single-screw man suggests that the mixing obtainable in multi-screws must be very limited because of the positive forwarding obtained in such machines.

Another difficulty occurs in considering the heating of multi-screw machines as against single screws, which has been commented on briefly above. Under normal working conditions a relatively large proportion of the heat in a single screw is contributed by energy from the drive motor, whereas in a multi-screw machine the major part is obtained by conduction from an external source. Extrusion experts have suggested that the former method is the most likely to produce a homogeneous melt, thus casting doubt on the effectiveness, as regards heating, of multi-screw machines. This is an opinion which is not always justified in practice.

It is obvious that because of their more complicated construction multi-screw machines will generally be more costly than a single-screw extruder of comparable output. Furthermore, due to thrust bearing limitations and drive complications such machines will be more difficult to maintain. Another important limitation of multi-screw systems results from the geometry of the adaptor section of the barrel, which renders effective streamlining of this zone extremely difficult.

In general, therefore, multi-screw extruders tend to be used only where these limitations are unimportant or where the extra costs are justified by some processing advantage.

Multi-screw machines will accept difficult feed materials more readily than will single screws, and will forward such materials to the die with less supervision of processing conditions. Because of this there is a growing trend towards the use of multi-screws for the continuous production of PVC compounds in place of internal mixers and towards the multi-screw extrusion of unplasticised PVC sheet and other products direct from a premixed powder.

Because of the different pressure mechanisms which exist in the two types of machine, the multi-screw extruder is less dependent on die restriction to build up back pressure in the system. Such machines therefore are more effective compared with single screws for the extrusion of heavy sections, particularly in difficult materials.

REFERENCES

1   FISHER, E.G., 'Trends in Extrusion Machinery', *Br. Plast. mould. Prod. Trader*, **26**, 392 (1953)
2   FOLLOW & BATE LTD, Manchester, private communication to the author
3   LEVER BROTHERS CO., U.S. Pat 2 048 286 (17 August 1933)
4   ANON, *The Extrusion of Thermoplastics*, R.H. Windsor, London (1950)
5   S.p.A. LAVORAZIONE MATERIE PLASTICHE, Br. Pat. 629 109 (13 September 1949) and others
6   PASQUETTI, C., Br. Pat. 677 945 (7 June 1950)
7   ETABLISSEMENTS JUMEX S.A., Br. Pat. 691 135 (27 December 1950)
8   WELDING ENGINEERS INC., U.S. Pat 2 441 222 (19 October 1943)
9   REIFENHAUSER, H. Br. Pat. 860 876 (19 July 1957)
10 MÄKELT, H., 'Observations on Machine Construction in the Development of Twin Screw Extruders', *Int. Plast. Engng*, 4, 20 (1964)
11 KRAFFE DE LAUBAREDE, L.M.H., Br. Pat. 676 325 (20 February 1950)
12 T.C. FAWCETT LTD, Br. Pat. 180 035 (16 May 1922)
13 ANON., 'The Mixtruder', *Rubb. Plast. Age*, **37**, (1956)
14 PASQUETTI, C., Br. Pat. 638 364 (2 April 1947)

5

Extruders for Special Purposes or of Unorthodox Design

5.1 Introduction

It is well known and has been stated many times in this volume and elsewhere that for maximum theoretical efficiency an extruder screw should be designed specially to suit the process and material with which it is to be used. Moreover it has been inferred that ideally a new design of screw could be specified for each different die/material combination.

Obviously the vast range of screws and other equipment, to say nothing of the frequent changes of setup which strict adherence to the above principle would entail, could quickly destroy the profitability of many present-day custom extrusion firms with their frequent short runs spread over a wide range of products. On the other hand a good case can generally be made out for the design of special equipment to handle a range of products of similar type with the best possible compromise efficiency and much of the theoretical work on extruder design has had this aim in view. The design of valving systems and of venting and the work carried out on short-transition-zone/long-metering-zone screws are examples of this which immediately come to mind.

In addition to the foregoing, however, there are always many other reasons such as product quality or even expediency, as against the desire to obtain the maximum extrudate throughput, to prompt the design of extrusion equipment to suit a particular need or range of products. The crosshead and the offset arrangements of die heads, which are described in detail in Chapter 8 of this work, are two of the many examples of particular extrusion systems which have evolved over the years to simplify the production of specific extruded products and do not aim, except perhaps indirectly, at improved outputs. Twin-screw extruders in their various forms can be considered in a similar light, as in fact can almost every die, calibrator and take-off arrangement, etc., now used in the industry.

Indeed it can be stated that the extrusion industry has been built up on a basis of expediency by men who, wanting to extrude a particular new product, or simplify the extrusion of an existing product, have constructed the equipment to do so by a process of trial and error. The theoretical workers have then come along afterwards to analyse the results and to suggest improvements.

The practical reasons or reasons of expediency which have prompted the design of such special equipment, or the modification of existing equipment, have frequently been those of space, convenience in handling, labour content related to the average size of order, cost of equipment and ease of changing such equipment for a new setup, etc., in addition to those of product quality and output. Other reasons have been the need for products of a special type not extrudable on existing combinations of equipment. Thus there is continual experimentation with new and perhaps unorthodox designs of extrusion equipment or with new combinations of old equipment. Many unusual arrangements have been produced, therefore, some to be accepted and become standard and others to be rejected and eventually forgotten. A few of the more important or more interesting of these are discussed below.

5.2 Extruders for large sections

It is generally recognised that for best results the area of a die orifice must be related to the size of the extruder with which it is used, irrespective of the rate of output in kilogrammes per hour (pounds per hour). As a guiding principle it is often stated that with modern equipment the die orifice area should never exceed half the cross section of the machine barrel bore and for best results the maximum ratio of 1 : 3 should be adhered to. A simple calculation based on these ratios shows that the maximum section which could be extruded on a 63 mm (2½ in) machine, irrespective of its efficiency, is a 75 mm (3 in) o.d. tube with a 6.3 mm (¼ inch) wall and from experience with such matters even this seems to be optimistic.

Multi-screw extruders, on the other hand, despite their comparatively low outputs in relation to screw diameter, are known to be very successful in the production of large-diameter pipes and other heavy sections which would be beyond the capacity of a single screw machine of comparable output. This anomaly results from the different pressure mechanisms which obtain in the two types of machine as discussed in Chapter 4.

As explained in Chapter 8, the extrusion die besides shaping the product to the required form must also offer a restriction to build up back-pressure in the screw system in order to work and homogenise the material. A large die, therefore, offering little restriction to flow, would result in a poorly homogenised extrudate with unfluxed granule inclusions unless backed up by some other restrictive system in addition to the die. Such restriction systems can consist of heavy screen packs or a controllable valve in the die adaptor.

However, even though such methods of restriction are very effective in producing a homogeneous melt, the feeding of a very large die—particularly a large-diameter thick-walled tube die—from a small extruder brings other problems. The large degree of radial flow necessary to fill the die annulus and the resulting extended residence time are important, as is the increased difficulty, due to this radial flow, of overcoming the velocity disturbances resulting from the mandrel support or spider structure.

Figure 5.1. 150 mm machine for large-section extrusion. (Courtesy Frieske & Hoepfner G.m.b.H., Erlangen-Bruck)

Multi-screw extruders are not die-sensitive in this way, which accounts for the greater success of machines of this type in large area extrusion. In a multi-screw machine the pressure mechanism is more self-contained and positive whilst the heating of the material occurs is largely by conduction. Thus it is often possible to obtain a properly homogenised output directly from the screws of such machines without a die or other restriction so that the filling of a large die annulus presents little difficulty.

The obvious answer to the problem of extruding large, heavy sections from a single-screw machine, however, would seem to be in the use of an extruder of adequate size. From the ratios previously stated a modern 200 mm (8 in) machine should be capable of producing an 450 mm (18 inch) o.d. pipe with a 9.5 mm (⅜ inch) wall without too much difficulty.

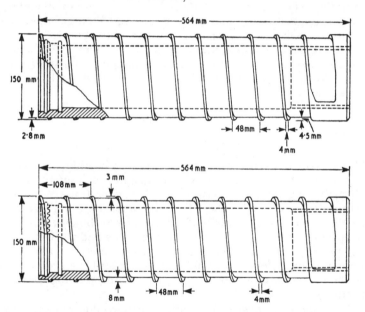

Figure 5.2. 150 mm tapered-compression-zone screw, and 150 mm short-compression-zone screw. (Courtesy Frieske & Hoepfner G.m.b.H., Erlangen-Bruck)

As always, however, even this solution is not perfect. The demand for large extrusions although improving, is not high either in terms of length of run or in tonnage output. It is doubtful, therefore, if a 200 mm (8 in) machine with the space it would occupy, the power it would consume, and the extensive ancillary equipment it would need to handle the high output, could be justified in any but the most exceptional of cases. Moreover, the high linear rate of throughput at which even a large section would come from a 200 mm (8 in) machine would be very difficult to handle and cool to ensure a perfect product.

It was for the above reasons that Messrs Frieseke & Hoepfner of Erlangen-Bruck produced their type SP150 machine to combine the advantages of a large-diameter screw with the low power consumption and floor space, low cost and ease of product handling of a small machine[1] (*Figure 5.1*).

Contrary to present day trends, this extruder was equipped with a 150-mm diameter screw of 3.5 : 1 L/D ratio. The theory behind the design is that the channel length and capacity of a screw of this diameter could be made to equal that of a normal 60 mm screw if the helix angle and rotational speed were suitably modified. As would be expected, therefore, the helix angle is very small and the screw speed is low, but in order to satisfy a wide range of requirements the makers were prepared to modify both of these factors to suit the characteristics of the material with which the machine was intended to be used (*Figure 5.2*). Although demonstrating an interesting concept these machines were not commercially successful and are no longer in production.

5.3 Vertical extruders

Many standard extrusion processes involve a change of direction of the melt between the screw and die and much attention has always been given to the design of die heads and flow channels to minimise the effects of these flow disturbances. Such direction changes occur for example in the extrusion of tubular film and in the production of parisons for blow moulding and whilst they do not present difficulties when easy-flowing materials such as the normal polyethylenes are used, the problem could become serious in the likely event of PVC materials becoming popular for these products.

The concept of a vertical extruder which would avoid direction changes in such processes has often been discussed and at least four firms are now offering equipment of this type. A further important advantage of vertical extruders is the saving of floor space they offer whilst a possible disadvantage is the lack of flexibility resulting from the machines becoming substantially single-purpose units.

Perhaps the earliest example of an interesting vertical extruder is described in a patent by the Chrysler Corporation of Delaware[2].

This extruder, which was intended as an injection unit for a moulding machine, consisted of a screw and barrel arranged to operate in a vertical position. The barrel was provided with two semi-circular recesses or housings on opposite sides, at a distance of at least one flight pitch of the screw from the feed position, in which a pair of gears were mounted to be in worm and gear engagement with the flights of the screw. The purpose of these gears was to prevent back flow and slip of material and to assist in the homogenisation thereof, the machine thus becoming in effect a form of multi-screw extruder.

This interesting machine is believed to have been used by the Chrysler Corporation for their own purposes and there is no record of it having been offered for sale or operated by companies outside the Chrysler Corporation.

Vertical extruders of a more conventional type—apart, of course, from their direction of operation—are now produced by Krauss-Maffei, Munich, Paul Troester, Hannover-Wulfel; and Oerlikon Plastic, Zürich, Barmag and others.

The Krauss-Maffei machine is intended for normal extrusion and possesses a screw of 10.5 : 1 L/D ratio equipped with an elongated conical fluted mixing or milling head which is presumably intended to overcome any loss of homogenisation due to the relatively short screw. The use of a vertical machine for normal extrusion seems to defeat the main object of this principle of operation in that it introduces a change of direction which apart from wire covering work does not exist with the normal horizontal machine. The advantage of space saving does remain, however.

The Paul Troester equipment (*Figure 5.3*), which is typical of many such machines, has been designed for attachment to cable and sheathing or insulating dies for longitudinal colour stripe marking.

Among the most interesting vertical machines at present available are those made by Oerlikon Plastics. These machines are outstanding in that they are single-purpose units which have been designed to take full advantage of the vertical system of operation in the manufacture of products to which this particular method is most suited.

The first machine produced by this firm forms the basis of an extrusion coating line. The extruder, which is available in 100 or 150 mm sizes, is suspended vertically downward and equipped with a coating die of substantially normal type with the coating equipment immediately under. The extruder may be moved, on the horizontal rails from which it is suspended, to or away from the coating position by motor drive and push-button control.

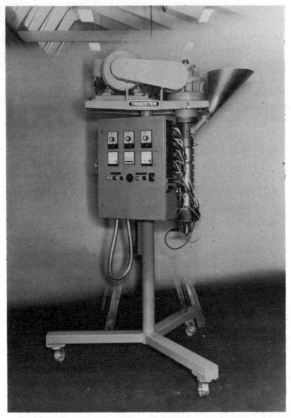

Figure 5.3. Vertical extruder for the stripe marking of cables. (Courtesy Paul Troester, Hannover-Wulfel)

This arrangement, therefore, besides avoiding the long 90° die attachment elbow which often gives trouble owing to stagnation and burning at the high temperature used in this operation, also provides easier access to the coating equipment than is normally available on horizontal setups.

Moreover, although extrusion coating and laminating is normally carried out with the polyolefins, equipment of the type described above fitted with a suitable die as discussed in Chapter 8 could be used with PVC and other heat-sensitive materials.

The second interesting Oerlikon machine is known as the Rotatruder and is designed solely for the production of tubular blown film. This machine, which at the time of writing is available in 40, 60 or 90 mm sizes, consists of a complete extruder with drive, heat

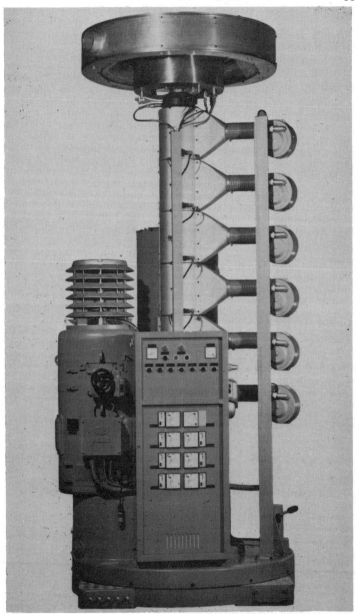

Figure 5.4. Vertical extruder for blown tubular film production, mounted on an oscillating base. (Courtesy Oerlikon Plastics, Zurich)

control instruments, die and tubular film cooling equipment, mounted vertically upwards on an oscillating base (*Figure 5.4*). By means of this oscillating base the small and inevitable variations in thickness of the tubular film are distributed evenly around the tube so that smooth tight rolls, which are very desirable in converting operations involving automatic bag making and printing equipment, are always produced.

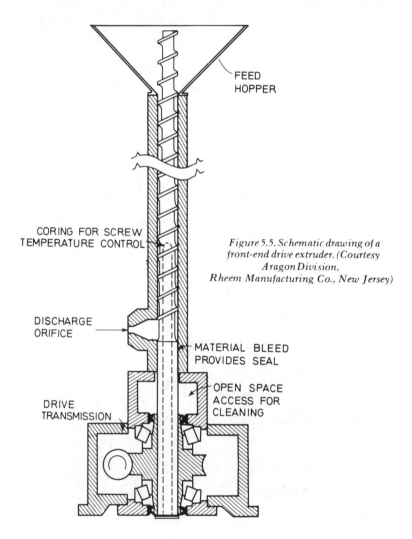

FEED HOPPER

CORING FOR SCREW TEMPERATURE CONTROL

DISCHARGE ORIFICE

MATERIAL BLEED PROVIDES SEAL

OPEN SPACE ACCESS FOR CLEANING

DRIVE TRANSMISSION

Figure 5.5. Schematic drawing of a front-end drive extruder. (Courtesy Aragon Division, Rheem Manufacturing Co., New Jersey)

The vertical extrusion process for tubular film, therefore, as typified by the Oerlikon Rotatruder, besides giving the advantages of limited floor space and direct die attachment without direction change, also provides this oscillating feature which is more complicated to duplicate in a conventional arrangement employing a horizontal machine.

An interesting departure from the normal screw drive systems for extruders is shown in the range of Aragon machines manufactured by Aragon Division, Rheem Manufacturing Company, New Jersey. In these extruders, which are also vertical but downward-pointing, the drive is positioned at the discharge end of the screw and the extrudate emerges at 90° to the screw and barrel axis. The advantages claimed for this unorthodox arrangement are that as the drive is taken by the thickest and therefore strongest part of the screw, higher torque can be applied and the feed section can be deeper, giving greater throughput capacity. Further claimed advantages are that the thrust bearing arrangements can be simpler and less costly and that the feed section of the screw can be extended into the vertical hopper to give a very simple force feed system for difficult materials. The Aragon machines have been widely used for the compounding of PVC gramophone record compounds; a schematic drawing showing such an arrangement is given in *Figure 5.5.*

A similar machine manufactured by Gloucester Engineering Co. Inc., Gloucester, Massachusetts and known as the Vertruder, is incorporated into a film scrap recovery unit where the extended screw, force-feeding the lightweight cut film enables a throughput of up to 295 kg/h (650 lb/h) to be obtained. A schematic arrangement of this unit is shown in *Figure 5.6.*

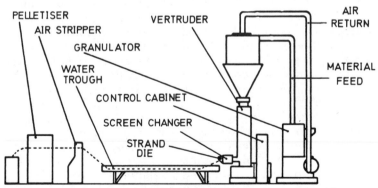

Figure 5.6. Schematic drawing of the Vertruder Model 4500. (Courtesy Gloucester Engineering Co. Inc., Gloucester, New Jersey)

5.4　Extruders for compounding and mixing

Mixing and compounding are highly important processes in the plastics industry. Most plastics materials must undergo some form of mixing operation before they are suitable for fabrication into end products and in many cases more than one mixing process is involved[3]. The preparation of a normal plasticised PVC compound, for example, involves first blending the polymer with the other dry ingredients such as heat stablisers, fillers, lubricants and pigments, and then incorporating the liquid plasticisers in the same equipment. The blend so formed is subsequently transferred to a high power internal mixer where, under the effects of intensive working, the polymer is fluxed to form a homogeneous mass with the other ingredients. The final step of the compounding process consists of discharging the mass onto a roll mill where it is sheeted into a suitable form for cutting into granules or cubes.

The compounding process described briefly above is the basic method used in the industry for the production of moulding and extrusion compounds. It is obviously a batch process which besides being unsound economically, also introduces the risk of contamination and of non-uniformity from batch to batch. Because of this there have been many attempts over the years to utilise the principle of the screw extruder to replace the internal mixer and thus to produce a semi-continuous compounding operation.

Screw extruders of both the single- or multi-screw types are known to be effective mixing devices so that at first sight the use of such machines for continuous compounding would seem to be a logical and easily realisable step. But, in fact, it has proved extremely difficult to develop a continuous screw compounder capable of producing compounds comparable in homogeneity and rate of output with those produced on modern internal mixers and it will be interesting to discuss briefly the reasons for this.

Mixing in polymer melts is obtained by shear and the effectiveness of the mixing action is determined by the rate of shear, the direction in which the shear is applied and the residence time. The dispersion of particulate matter in a polymer melt depends on the breaking down of the particles to the correct size and the random distribution of these particles in the melt. The melt flow in a screw extruder is laminar, therefore a material fed to such a screw will be sheared into layers which if sufficiently thin will break down the particles to the required size for a good mix. The thinness of the layers is again dependent on the rate of shear and on the residence time, which in turn are dependent on the restriction in the system.

So far with a screw extruder of substantially normal design it has been found impossible to introduce sufficient shear into the system to break down particulate matter to the required fineness without increasing the restriction and thus the residence time to an uneconomic output level. Moreover, the laminar flow in an unmodified screw machine is substantially circumferential so that even with a high degree of restriction the extrudate tends to be circumferentially striated[4,5].

In order to overcome these difficulties, therefore, most of the continuous compounders now available aim at applying additional but localised intense shear in a direction perpendicular to the normal laminar flow of the melt. Many interesting devices have been tried for this application and the patent literature is so extensive and complicated that any attempt to cover the subject completely would require its own book.

Continuous screw mixers can be divided into two main groups, i.e. substantially normal extruders which are fitted with ancillary attachments for mixing and machines which have been developed from screw extruders but which are suitable only for mixing.

5.4.1 EXTRUDERS WITH MIXING ATTACHMENTS

One of the best known examples of the first group of machines is due to Frank Egan & Co. of the U.S.A., and consists of a planetary mixing head fitted to a normal extruder. As seen from *Figure 5.7*, this head contains a central pinion attached to the extruder screw; this drives a series of planet gears which in turn are in mesh with an internally cut gear ring attached to the main housing. In operation the planet wheels are driven by the central gear and the plastic material passing through the head is subjected to considerable shear in a direction perpendicular to the laminar flow in the screw and yet is not severely restricted. In addition to the intense shear produced by this device there is also a pronounced degree of metal to metal smearing and rolling between the gear teeth which must be of considerable importance in the breakdown of solid particles. The German company Schalker Eisenhutte, now Eickhoff, have in effect reversed the Egan principle by installing the planetary device at the feed end of the extruder or in the transition zone. It is claimed that this arrangement gives a positive multi-screw feed with improved mixing which can be operated in the normal manner.

Although the author has not seen this machine in operation it could be suitable for rigid PVC extrusion direct from powder blendes (*Figure 5.8*).

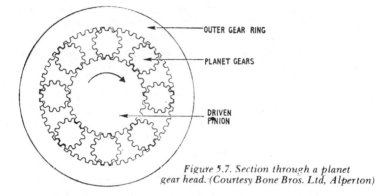

Figure 5.7. Section through a planet gear head. (Courtesy Bone Bros. Ltd, Alperton)

Perhaps the earliest machine to be equipped with a mixing attachment is the Gordon Plasticator manufactured by the Farrel (Birmingham) Co. Inc., which although originally designed for the compounding of rubber has also been adapted to the production of compounds in PVC and other materials. This machine is basically a normal extruder, but with a fluted conical nose fitted to the end of the screw. The screw also contains a central milling section, but it is doubtful if this contributes at all to the mixing action of the system.

The fluted conical nose is arranged to engage with a smooth female cone in the head and the clearance between these two members is adjustable by means of a series of interchangeable distance pieces of different thickness. The Gordon Plasticator is thus one of the earliest machines to use adjustable restriction in the head.

Two further interesting machines which can be considered in this group are the Andouart Gellifier and the Pastorello Mixer/Extruder.

The Andouart Gellifier was basically an extruder of normal type but with the screw divided into two or more sections as with a normal vented machine. The sections of the screw were separated from each other by interchangeable semi-circular screen plates which passed through the barrel walls on opposite sides to engage with appropriately positioned plane portions of the screw shaft. A vacuum extraction device was fitted immediately downstream of the divided screen plate and the machine was in effect a vented extruder in which the degree of restriction prior to the decompression zone could be adjusted by changing the screen plates (*Figure 5.9*).

The Pastorello machine combines a mixer of the ribbon blender type with an extruder of substantially normal design, the purpose

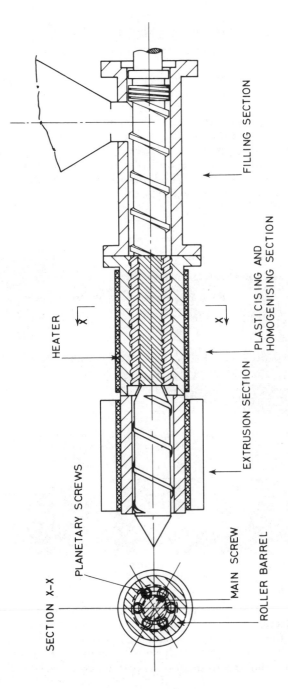

SECTION X-X

PLANETARY SCREWS

MAIN SCREW

ROLLER BARREL

HEATER

FILLING SECTION

PLASTICISING AND
HOMOGENISING SECTION

EXTRUSION SECTION

Figure 5.8. Diagrammatic illustration of a planetary screw extruder. (Courtesy Firma Eickhoff, Germany)

being to obtain a finished extruded product directly from uncompounded ingredients. Essentially the equipment consists of a stainless steel jacketted blender with a single-screw extruder protruding from one end in line with and turning on the same axis as the blender agitators. The screw may be driven in either direction independently of the blender agitator blades and in operation the ingredients are first charged to the blender vessel with the screw rotating in the non-feed direction. On completing the mixing operation the direction of screw rotation is reversed to enable the screw to feed. Thus the operation of this equipment would seem to be of the batch type, first to prepare a compound in the form of a dry blend and then to extrude immediately the finished product.

Figure 5.9. The Andouart Gellefier. (Courtesy Societe des Etablissements Andouart, Bezon, Paris)

Figure 5.10. The Ross ISG Mixer. (Courtesy Dow Chemical Company)

Further interesting mixing attachments for extruders are the static mixers offered by Kenics Corpn and the I.S.G. unit developed by the Dow Chemical Company. This latter equipment is a continuous pipeline mixing head with no moving parts but containing a series of tetrahedron-shaped members each of which is pierced with four bores which connect the tetrahedra to each other. The material passing through the bores in the tetrahedron-shaped members is minutely divided and dispersed, giving a high mixing effect. The body of the attachment can be of any convenient length to suit material requirements[7]. The Kenics unit is similar but uses a different design of dispersion head[8]. Both units are claimed to give increased production, better mixing and a lower-temperature extrudate.

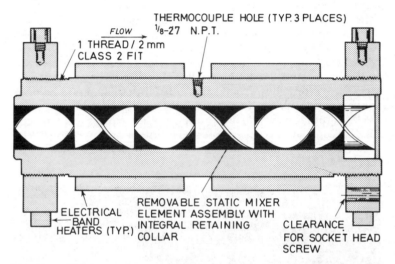

Figure 5.11. Schematic illustration of the Kenics Thermogenizer. (Courtesy Kenics Corpn, Danvers, Massachusetts)

Both the I.S.G. head and the Kenics unit are illustrated in *Figures 5.10* and *5.11* respectively.

5.4.2   MIXING MACHINES BASED ON SCREW EXTRUDERS

The machines described in the previous section are normal extruders with mixing attachments and if necessary they could be used without substantial change for the production of ordinary tube, profiles, etc. The machines of the second group now to be described have been developed specially for mixing or for continuous compounding and could not easily be used for normal extrusion operations.

One of the best known machines in this group is the Buss Ko-Kneader which is of novel and unique design. This machine has a screw with slotted flights which besides rotating also reciprocates axially. The barrel contains three longitudinal rows of inwardly projecting pegs which engage with the slots in the screw flights during the reciprocation/rotation of this member to transfer material from one flight to the next. This action tends to break up the flow pattern of the screw and to give intense shear in a direction perpendicular to the normal laminar flow.

A second interesting mixing machine which departs from the normal action of the screw extruder is the Werner & Pfleiderer Plastificator. The basic element of this machine is a fluted tapered rotor revolving in a non-fluted stator. The clearance between these two members is finely adjustable during operation so that the amount of shear imparted to the material can be accurately controlled. Pre-blended material is fed to the rotor by means of a twin-screw conveyor and is sheared and rolled into filaments of homogenous compound. The output from the rotor passes directly into a normal single-screw extruder built on to the equipment whereby the filaments are finally compounded into a homogeneous mass before feeding to a die face cutter situated at the extruder head (*Figure 5.12*).

Figure 5.12. Plastificator PK 2000 II with ancillary granulating equipment and fluidised-bed pellet cooler. (Courtesy Werner & Pfleiderer, Stuttgart)

A special machine, the Kombiplast, also manufactured by Werner & Pfleiderer, for the compounding of PVC has separated the three stages in the process giving each a specifically designed processing unit. The first of these is the feed section where a single screw conveyor gives a positive and consistent feed to the second unit. This is a high-speed, high-shear, twin-screw compounding section using screws of special design combined with kneading blocks to provide controlled shear and give maximum compounding and pigment dispersion. At the end of this section the material in gel form is fed into the final unit which is slow-speed, single-screw discharge extruder with a variable-speed die face cutter. Volatiles are removed from the compound as it is transferred to the discharge extruder.

The British-made Extramix manufactured by Christchurch Engineering Co. Ltd, Parkstone, Dorset, also uses the flighted rotor system revolving within a barrel which is fluted to provide high shear forces on the material. In this machine helical grooves of opposite hand are cut in the rotor and the barrel, the barrel grooves increasing in depth as the rotor grooves get shallower and vice versa. The machine is used extensively for compounding mixed polymers and fibrous polymer mixes. In appearance it is similar to a standard extruder except that the L/D ratio usually is less and the rotor speed is higher.

5.4.3 MULTI-SCREW MACHINES

The screw mixers which have so far been described have all been based on the single-screw principle of operation. There have been many developments of the multi-screw principle, also, and some of the most successful continuous compounders have been based on this system as briefly mentioned in Chapter 4.

The multi-screw extruder, however, is capable of such a vast range of variations in screw profiles, barrel and adaptor geometry, feeding heating and restriction systems, not to mention interscrew clearance and relative direction of rotation, all of which have an effect on the extrusion and mixing efficiency, that it would be impossible to describe each maker's interpretation adequately in this book.

One of the best-known continuous twin screw compounders is the machine produced by Welding Engineers Inc. of Norristown, Pennsylvania. This machine, which is made in a range of sizes up to 150 mm (6 in) screw diameter, has two non-intermeshing screws which rotate in opposite directions in a figure-of-eight bore. The spacing of the screw centres is equal to their diameter and both screws rotate at the same speed which may be up to 150 rev/min or more depending on the application.

128

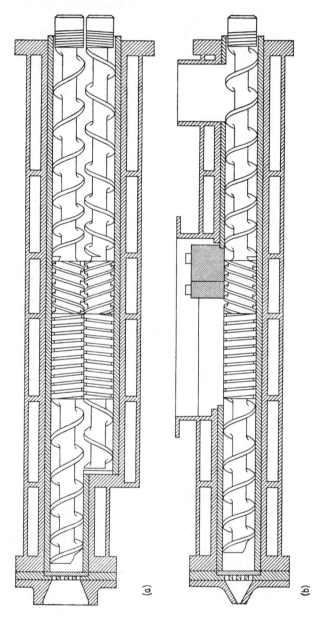

(a)

(b)

Figure 5.13. Schematic arrangement showing screws of the Welding Engineers System. (Welding Engineers Inc., Norristown, U.S.A.)

The design of the Welding Engineers machine is unique in that it uses two screws which are of different lengths, giving a twin-screw feeding zone and a single-screw discharge, and because the screws may both contain short sections of flights cut in the reverse hand.

The reverse flights act as a braking zone and give a definite back flow to improve the mixing effect. Because the reverse flights are cut with a smaller helix angle compared with that of the feed and discharge zones, the overall forward movement is maintained and the geometry of this reverse section can be changed to suit the viscosity and other characteristics of the material being processed[5] (*Figures 5.13* and *5.14*).

Figure 5.14. 114 mm (4½ in) Welding Engineers twin-screw compounder. (Courtesy Welding Engineers Inc., Norristown, U.S.A.)

The second twin-screw compounding system to be described is that of the Krauss-Maffei DSM 11/150. This machine is unique in that it aims at combining the functions of an internal mixer with those of an intermeshing twin-screw extruder and a pair of substantially normal single screws.

Thus the feed zone of this machine consists of a pair of large-diameter intermeshing flight screws operating in a figure-of-eight bore. Immediately downstream of the feeding zone is situated a pair of intermeshing kneaders leading into conical compression zones. The compression zones, which are adjustable by axial screw movements, guide and pressurise the material into a pair of separate single screws, each working in its own individual bore, which are continuous with the kneading and feed zones (*Figure 5.15*).

The third machine to be described is not strictly a compounder because it is recommended by its maker for the direct extrusion of materials in the powder form, at which function it would seem to have considerable possibilities of success.

This chapter, however, is intended to deal with extruders of unorthodox design and as the machine under discussion could also form the basis of a continuous compounder it seems relevant to discuss it in this section.

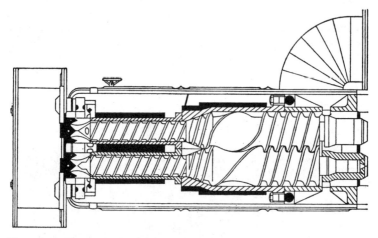

Figure 5.15. Sketch showing the arrangement of the screws of the Krauss-Maffei DSM 11/150 machine. (Courtesy Krauss-Maffei, Munich)

The machine therefore is the Berstorff types KD120 (and 90) already referred to in Chapter 4, which is unique in that it possesses a short intermeshing conical twin-screw section feeding a long single screw of normal design which is continuous with the main screw of the feeding section (*Figure 4.8*). As with all normal single-screw machines various designs of main screw are available from the makers—including one with an eccentric root—and this equipment would seem to combine the excellent feeding and pressurising characteristics of the twin screw with the superior screw and adaptor geometry and design simplicity of the single.

The last continuous compounder to be described in any detail is the Type ZSK 83/700 by Werner & Pfleiderer. This machine possesses several novel features and is designed in such a way that the flight and kneader configuration of the two screws can be changed to suit individual processing requirements.

In general design the ZSK 83/700 consists of an intermeshed twin-screw machine in which the screw flights and other interacting members are mounted and locked on parallel shafts so that they can be dismantled and replaced. The kneading sections consist of triangular-shaped discs, whose number may be varied to meet specific requirements, which are assembled on the shafts between the feed and delivery screw sections. Different screw lengths are available to accommodate various kneader assemblies (*Figure 5.16*).

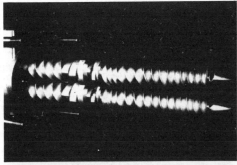

Figure 5.16. Above : one of the many screw/kneader disc arrangement possible with the Werner & Pfleiderer ZSK 83/700 twin-screw compounder; below: the same machine showing screw assembly. (Courtesy Werner & Pfleiderer, Stuttgart)

Werner and Pfleiderer have also produced a four-screw machine, model VDS-V, with the screws arranged in two pairs to form a V configuration. Each pair of shafts are co-rotating and have a self-sealing profile similar to that of the twin-screw machine. The same principles of variable-geometry screw and barrel sections are applied. The V form of design offers several advantages, particularly in high devolatilisation processes such as the continuous concentration of atactic polypropylene. A diagram of the VDS-V is shown in *Figure 5.17*.

Before leaving this question of novel continuous compounders it is perhaps relevant to refer again to the Pasquetti Bihelicoidial machine which has been described in Chapter 4 and illustrated in *Figure 4.11*. Although there are certain obvious limitations to a

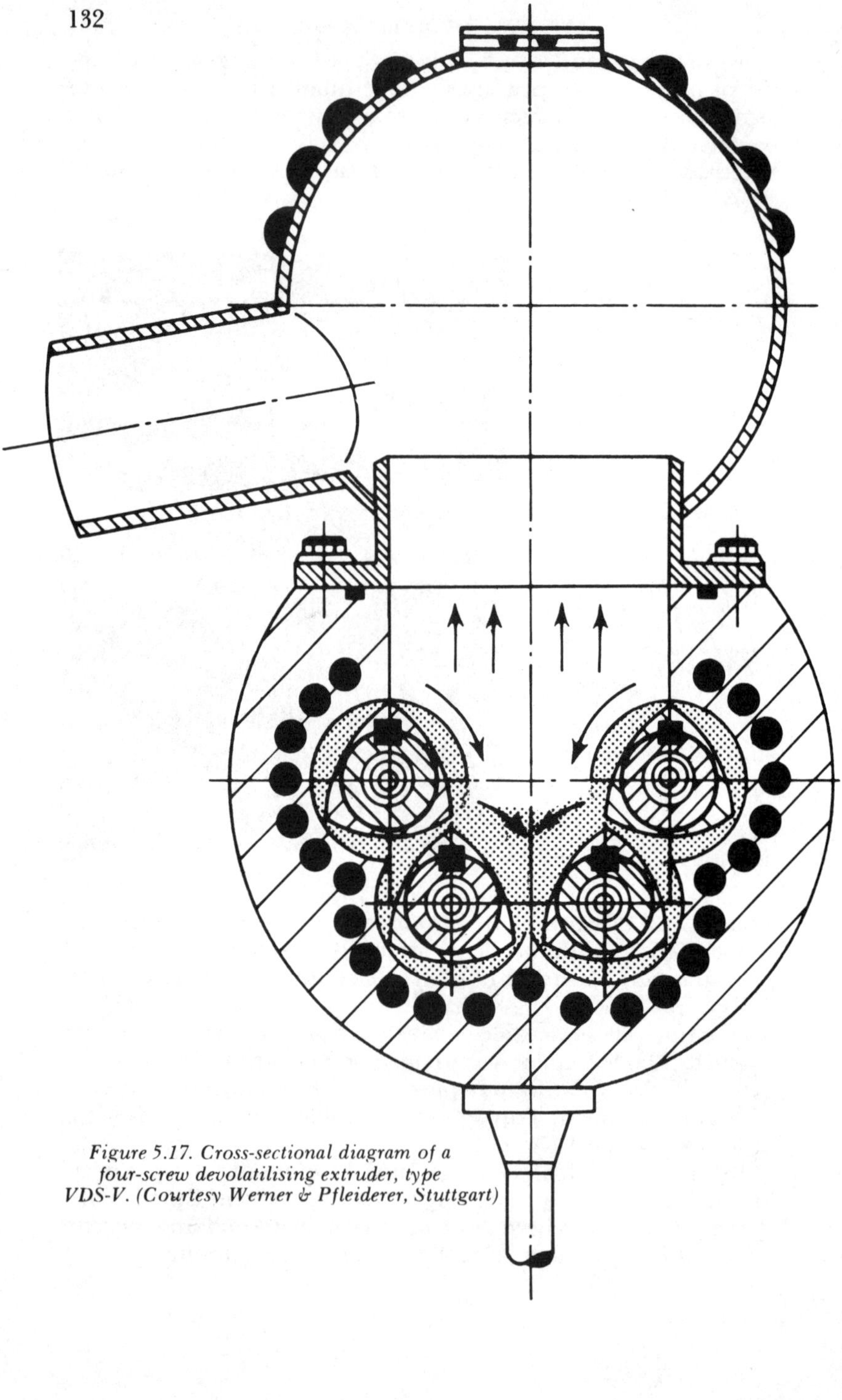

*Figure 5.17. Cross-sectional diagram of a
four-screw devolatilising extruder, type
VDS-V. (Courtesy Werner & Pfleiderer, Stuttgart)*

mechanism of this type such as the short residence time and the difficulty of providing lubrication to the gear shaft bearings, the principle is interesting in that it is an attempt to provide a form of continuous milling in a small space and with a minimum material inventory. It seems rather strange, therefore, that this principle has not been further utilised or investigated.

5.5   Extruders of unorthodox design

In Section 5.1 it was stated that many unusual extruders have been developed over the years and this is obviously true. The patent literature abounds with strange devices for extrusion, many of which have never been realised, so far as the author is aware, in a practical commercial form whilst others which have shown great promise were never taken up and further developed. There have been extruders with helically grooved and rotating barrels, extruders consisting of one or more pairs of interconnected screws operating in a horizontal opposed fashion feeding to one centrally disposed crosshead and extruders wherein the screw thread has in effect been cut on the face or faces of a rotating disc.

Extruders have also been developed consisting of conical screw members rotating eccentrically within stationary housings having female threads of similar form. Others have been constructed having no forwarding flight formation at all, but relying solely on shear forces of a special type.

Apart from complete extrusion machines which have been used, or attempted to use, new principles of operation, there have also been many departures from the normal conception in the design of screws and other details for inclusion in machines of otherwise orthodox construction.

A very few of the more important or interesting of these innovations will be described.

5.5.1   THE MAILLEFER AND OTHER DOUBLE-CHANNEL SCREW SYSTEMS

A recent and extremely interesting development in extruder screw design, which has already been referred to in earlier chapters, is due to Dr Charles Maillefer of Switzerland, who although an expert in the generally accepted concept of extruder screw flow mechanisms[10] was of the opinion that these theories were limited in practical application.

Maillefer therefore considered carefully the state of the material during its transition to a melt and produced a screw of unorthodox design to improve the heat transfer.

This screw consists of two separate channels cut in the same shaft, one of which connects with the feed hopper whilst the other channel supplies the melt to the die. The feed channel width decreases progressively until it disappears entirely at a point about two thirds of the way along the screw length and this channel may be regarded as feed and part transition zones. The width of the output channel increases progressively from zero at a point just downstream of the feed hopper and attains full width at a point adjacent to the position at which the feed channel width is zero. The only connection between the two channels is over the flight lands, which can perhaps be regarded as the true transition zone. The viscous melt which forms in the feed channel in contact with the barrel is transferred to the output channel over the flight land so that in effect one channel is filled with fully melted material and the other with material in either granule or transition form. The heat transfer from the screw and barrel wall is, therefore, much increased and all material reaching the output channel will have been subjected to substantially similar heating and shear effects. The makers claim that better quality products at higher outputs per unit of power are obtained[11] (*Figure 5.18*).

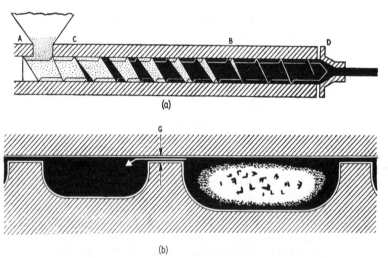

(a)

(b)

Figure 5.18. Diagrammatic arrangement of the Maillefer screw showing (a) the flight arrangement, and (b) the transfer of the material between channels. (Courtesy Maillefer S.A., Renens-Lausanne)

No precise measurements of the screw are given but it seems probable that the gap between the flight lands and the barrel is larger than in normal machines. Even so, there must be a very high degree of shear in this area and difficulty in processing rigid PVC, acrylics and other heat-sensitive materials might occur.

Many other interesting special purpose screws have been designed, including of course the barrier-type mixing screws which were mentioned in an early chapter of this book.

Another interesting design is the 'Barr' screw, so named after the work carried out by R.A. Barr and C.I. Chung of the Waldron-Hartig Division, Midland Ross Corporation, which is somewhat similar to the Maillefer screw in that two channels are used. However, instead of progressively varying the width of the channels to compress the material, as adopted by Maillefer, only the channel depths are changed. The material passes first to a single-channel feed section, the depth of which decreases to zero, where it is compacted and progressively melted at the surface by contact with the heated barrel. The molten layer is then forced by the screw pumping action over the flight lands into a second channel—the output section—which is of progressively increasing depth, and thence to the die system.

This design of screw is said to give lower melt temperatures compared with screws of orthodox design and obviously any machine in which it is incorporated can be used for any normal extrusion operation. A second stage can be fitted, for example for metering or to give a venting facility, or a compounding attachment can be added.

Modifications to this design were developed by the Hartig Corporation by the inclusion of passages in the feed channel communicating with the screw bore. The pressure generated by the solid material in the feed channel forces the melt through these passages into the screw bore and thence to the die. In addition, an internal stationary screw with a reverse helix profile is supported from the die end of the barrel to assist melt flow and to avoid 'hang-ups'.

Although ingenious in concept, the Hartig design suggested above was found to have a number of limitations in practical operation and modifications have been suggested. In one of these the melt was forced into a narrow but deep channel at the upstream side of the screw flight rather than into the screw bore.

In work carried out at I.K.V. Aachen[12] an analysis has been made of the function of the various parts of the conventional extruder screw. From this work it appears that the feed section operates at low efficiency. i.e. 20-40 per cent, and that there is a delay in the pressure build-up which requires the use of longer screws and also causes pulsating which results in poor uniformity in the end products. As a result of this a new design of feed section was suggested in which a tapered entrance bush with axial grooves was fitted. This system resulted in much more efficient conveying of granular and powdered materials and a higher quality product. In addition to this an oscillating coaxial ring plunger was used to force the material along the screw length to back up the action of the grooves in the bush. To prevent the melting of polymer prior to its arrival in the transition section of the screw the grooved bush in the feed section was isolated from the main barrel and cooled.

In the transition zone, Kosel suggests that the initial melting of the polymer forms pools of material at both edges of the flight, at the thrust edge by reason of the wiping effect of the flight and at the trailing edge from leakage flow, and he says that this leads to the development of a solid core of compacted material surrounded by a layer of melt with a consequent drop in efficiency of the overall melting process. It is then suggested that this partially homogenised material should be fed into a special mixing section of the screw. The transition zone of the screw would then be used only to stabilise the conveying action of the feed section and to prepare the initial melting of the polymer, the final melting being carried out only in the mixing section. By frictional working therefore, a plug of polymer is obtained which has a suitable melt exterior but because of the poor heat transfer properties of the polymer has a core of only partially melted material. To make this plug of material completely homogenised, therefore, a system whereby it is broken up by the action of the screw is necessary.

This is accomplished by changing the mixing section of the screw completely and replacing it with a series of slotted disc sections. For additional mixing the barrel section here has a series of fixed protuberances which fit between the slotted discs. The frictional heat developed in this section is quite high and any material not melted is transformed by shear and frictional heat. This system gives a very high degree of mixing and shear but offers a considerable restriction to the flow of material and in consequence reduces the output of the extruder.

5.5.2 THE ENGEL MACHINE

T.P. Engel of Offenbach/Heusenstamm has produced an extruder in which the material is heated to a relatively high temperature before being fed to the screw.

In essentials this machine consists of a heated horizontally rotating disc, equipped with a feed hopper and a scraper, associated with an extruder screw and barrel of substantially conventional design which is mounted diametrically above the disc with its feed opening downwards.

During operation, cold material is fed at a controlled rate to the heated rotating disc and after travelling through an angle approaching 360° is picked up by the scraper which guides it to the screw of the extruder. A speed of 6 rev/min for the disc gives a polymer residence time of less than 10 s before it enters the metering extruder. The final temperature at which the material enters the extruder is determined by the rate of feed from the hopper, which may be of the vibrating form, the temperature of the disc and its speed of rotation[13,14].

The author has had no personal experience of the operation or performance of the Engel machine, but it would seem to have outstanding possibilities with a number of thermally stable polymers.

5.5.3 EXTRUDERS WITHOUT SCREWS

Although in the plastics industry it is normal to regard an extruder as being exclusively a screw mechanism it is possible to use other systems of pumping to supply a plastics melt to a die. Ram systems may be used and gear pumps of various types are used in the spinning and other industries as previously discussed.

Screw extruders, however, are called upon to perform several other functions simultaneously in addition to supplying the material to the die. They are required to supply heat to the material both by conduction and shear, to introduce a mixing effect and to forward the material under controlled pressure conditions. Ram systems cannot perform these several operations efficiently because of the difficulties encountered in heating and because of the total lack of mixing effect. Gear pumps, although perhaps basically more suitable are not generally available in an appropriate form apart from the one example previously described (Section 4.3).

However, in October 1959, an extremely interesting article appeared in an American journal[15] describing an extruder which works without a screw and yet also performs the functions of heating, mixing and forwarding the material simultaneously.

This machine, which in essentials consists of a disc rotating in a stationary cup-shaped housing with a die orifice at its centre, functions by virtue of the centripetal pumping effect which results from the rotary shearing of a viscoelastic fluid in the gap between the face of the disc and its housing. The elastic recovery of such fluids when sheared along a curved path causes a force to be developed towards the centre of rotation and it is this force which enables the device to be used as an extruder.

The authors of the above article have constructed various laboratory models of the apparatus and have operated these models on normal extrusion processes using a range of thermoplastics materials. They have also carried out a series of tests on the apparatus and have produced a number of interesting plots relating output to temperature, speed of rotation of the disc, horsepower, and gap dimensions. The authors suggest that extruders of this type would be simple to operate, would have substantially reduced residence time compared with a normal screw machine, would give excellent mixing and be free from pulsating effects. The system would also seem to have possibilities in the cable industry, where the wire or other member to be covered could be easily passed through the centre of the disc and thus to the extrusion die, and the possibility of using such a machine as a pre-plasticiser for an injection moulder has been discussed elsewhere[16].

A further development in the search for a screwless extruder was announced[17] by engineers of the Bell Telephone Co. in the U.S.A. in 1962. Basically this machine, which is still in the very-early stages of assessment, comprises a cylindrical housing in which is mounted a stationary pressure plate, this plate having recesses cut in alternative segments around the inner face. Revolving in close proximity to this recessed plate is a flat disc. Plastics material is fed into the space between these two members and is continuously sheared by the action of the rotating disc. It has been found that by locating a series of holes in a stationary plate along the radii of the segments at their lowest point, a pressurised extrudate may be obtained and conveyed to a suitable placed die. It is claimed that this extruder has advantages over the elastic melt extruder when operating on low viscosity melts.

One of the disadvantages of this hydrodynamic extruder is the lack of pressure behind the material as it is extruded through the die. To overcome this a combined standard screw extruder and hydrodynamic unit has been made. In this the hydrodynamic unit—or as it is sometimes referred to, the 'elastodynamic' unit—is used to melt the material which is then fed into a standard extruder screw of short length. This then builds a pressure and forces the material through the die as in the conventional machine.

Diagrammatic cross sections of both these screwless extruders are shown in *Figure 5.19*.

A further unconventional extruder developed by Bell Telephone Laboratories is a system for obtaining a continuous extrudate from a ram extruder. To do this two ram-type injection moulding barrels feed into the same die and a shuttle valve arrangement permits the one to recharge the barrel while the other is extruding through the die. The advantages of this are claimed to be a much lower rate of shear of the material and a control over the maximum melt temperature, which is lower than with screw extrusion.

A two-roll mill system has been devised[18] which in addition to melting the polymer materials efficiently also generates sufficient pressure to extrude them through a die. The machine is mechanically simple and its performance is comparable with that of an equivalent screw extruder. One of its main advantages is the ability to accept a wide range of feed stock both in form and type. In this connection it is interesting to note that Iddon Bros in 1895 or thereabouts, also produced a two-roll mill system extruder for rubber.

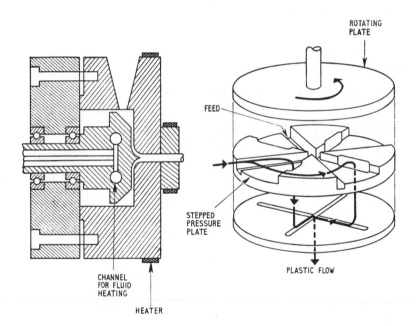

Figure 5.19. Diagrammatic cross sections of two suggested screwless extruders: (a) elastic melt extruder by Maxwell & Scalora (Princetown University, U.S.A.), and (b) shear disc extruder by Westover (Bell Telephone Laboratories, U.S.A.)

REFERENCES

1   ANON., 'The Frieske & Hoepfner SP150 Extruder', *Plastics, Lond.*, **25**, 66 (1960)
2   CHRYSLER CORPN, U.S.A., Br. Pat. 620 652 (28 March 1949)
3   FISHER, E.G. and CHARD, E.D., 'Principles of Mixing', *Int. Plast. Engng*, **2**, 54 (1962)
4   MOHR, W.D., SAXTON, R.L. and JEPSON, C.H., 'Mixing in Laminar Flow Systems', *Ind. Engng Chem.*, **49**, 1857 (1957)
5   MOHR, W.D., SAXTON, R.L. and JEPSON, C.H., 'Theory of Mixing in the Single Screw Extruder', *Ind. Engng Chem.*, **49**, 1855 (1957)
6   GEWERKSCHAFT SCHALKER EISENHUTTE, W. Germany, Br. Pat. 902 513 (7 December 1960)
7   DOW CHEMICAL CO., U.S. Pat. 3 404 869 (18 July 1966)
8   KENICS CORPN, U.S. Pats 3 286 992, 3 664 638 (24 February 1970) 3 704 006 (25 January 1971)
9   FULLER, L.J., Welding Engineers Inc., U.S. Pat. 2 441 222 (11 May 1948)
10 MAILLEFER, C., 'Etude theoretique et experimentale sur le Fonctionnement des boudineuses', *These de Doctrat*, Lausanne (1952); abridged form 'Analytical Study of the Single Screw Extruder', *Br. Plast.*, **27**, 394 (1954)·
11 ANON., 'New Extruder of Swiss Design', *Int. Plast. Engng*, **2**, 19 (1962)
12 KOSEL, U.M., 'A Novel Concept of Single Screw Extrusion', *Plast. Polym.*, **39**, 319 (1971)
13 ENGEL, T.P., Br. Pat. 873 188 (2 June 1959)
14 ENGEL, T.P., 'The Rotomeltor (Melt-Plate Extruder)', *Plastics, Lond.*, **28**, 72 (1963)
15 MAXWELL, B. and SCALORA, A.J., The Elastic Melt Extruder Works Without Screw', *Mod. Plast.*, **37**, 107 (1959)
16 FISHER, E.G. and MASLEN, W.A., 'Problems and Trends in Injection Moulding', *Br. Plast.*; **33**, 276 (1960)
17 ANON., 'New Approach to Screwless Extruders', *Mod. Plast.*, **40**, 119 (1962); WESTOVER, R.F., 'A Hydrodynamic Screwless Extruder', *S.P.E. Jl*, **18**, 1473 (1962)
18 POCKLINGTON, A.R., FENNER, R.T. and WILLIAMS, J.G., 'Extruding Plastics with a Two Roll Mill', *Plast. Polym.*, **41**, 253 (1973)

6

Constructional Features
of Screw Extruders

6.1 Introduction

The other chapters in this monograph have dealt, or will deal, in
some detail with the theory, design and operational features of
extrusion machines and their associated equipment but have commented only briefly on the more practical aspects of machine construction.

It is the aim of this chapter, therefore, to dissect the screw
extruder into its special component parts and to outline the
important constructional features of such parts.

6.2 Screw

Depending on its design an extruder screw consists of a shallow
cut zone at the discharge end with the root dimension tapering
down to a more deeply cut zone at the feed position and thence to a
parallel plain shank where it picks up the drive by means of one or
more keyways or splines.

An extruder screw may be subjected to very high torque loadings under working conditions and the total load must be transmitted by the reduced cross section at the feed zone which may be
even further weakened by a bore for water cooling. It is necessary,
therefore, to use a high tensile steel in order to withstand these
high loads. The commonest materials used for the manufacture of
screws are 750-950 MN/m² (50-60 tonf/in²) alloy steels, nitriding
steels of similar strength or stainless steels.

In addition to a high core strength, extruder screws also need to

141

have hard surfaces on the tops of the flights to prevent wear and to be resistant to the corrosive effects of the materials being processed or of their decomposition products. It is possible to obtain hard surfaces on the lands of screws made of most alloy steels by flame hardening, and nitriding steels can be treated to give a Diamond Pyramid figure of up to 1000. Difficulties arise when a stainless steel is used, however, and in such cases it is normal to use Stellite or some similar hard facing alloy along the flight land in the form of a continuous weld line and to grind to shape and size in a finishing operation.

Screws for PVC materials are often chromium-plated to prevent corrosion, but there is a certain divergence of opinion on the advisability of this practice. It is not generally considered correct to cover the flight lands with chromium plating, which is ground off in a final sizing operation. Difficulties can arise at the edge of the plating, at the flight land corner, which is prone to peel or chip. In cases where the possibility of corrosion exists, therefore, it is usually desirable to adopt the stainless steel screw with special alloy flight. Screws for polyvinylidene copolymers such as Saran are normally produced from special nickel alloys.

The means by which the screw is driven obviously depends on the overall design of the extruder. Some screws are constructed with an extended shank which passes right through the reduction gear into which it may be splined or keyed. This construction, although simplifying the water-circulating arrangements, presents machining problems and is not often used with the present day long screws. A more suitable method is to drive the screw by some form of splined or keyed coupling located on the hopper side of the reduction gear with a separate pipe screwed into the screw end to contain the cooling water return. Some makers use a flexible coupling at this point to insulate the screw from the reduction gear.

6.2.1 SCREW EXTRACTION

When an extruder is used on sticky melts such as polyethylene, cellulose acetate, etc., it often becomes gummed up in the liner and difficult to remove even when hot. Heavy screws on large machines are always difficult to extract, even when clean, owing to their weight. An important piece of equipment therefore which should be available for every extruder is the screw extractor. This consists of a crosspiece which can be attached to some suitably rigid part of the extruder body, usually on the rear side of the reduction gear housing, and a central nut in which a long screw can be turned manually to push the screw from the barrel. This piece of equipment should be robustly constructed so that the pushing screw does not bend under pressure and the end of the screw should be equipped with a pressure pad which will not damage the extruder screw end. The pushing screw should also be long enough to free the extruder screw completely under all conditions.

6.3 Barrel

The barrel of an extruder is the parallel cylindrical chamber in which the screw rotates and forwards the thermoplastics material. Early machines were constructed with barrels in sections to facilitate manufacture but modern machines are equipped with one-piece barrels which avoid leakage of material and overcome the problem of misalignment between sections.

For convenience of description, however, the barrel, like the screw, is still considered to be composed of three sections.

6.3.1 FEED SECTION

This section, as its name implies, is that part of the extruder barrel in which the material is fed to the screw, or screws, of the machine. There are two basic requirements for this section; firstly, the feed opening should be designed to permit the feed material to flow freely from the hopper, or feed device, into the screws without the possibility of bridging; and secondly, facilities should be available for water-cooling the section.

Early plastics extruders, being adaptations of rubber machines, usually had the feed opening arranged tangentially to the screw or even undercutting it. This arrangement made it difficult to fit a hopper or vibrator for the feeding of granular material, and increased the cleaning troubles as well as giving unequal feed rates. Machines for thermoplastics were, therefore, soon constructed with feed ports opening radially into the top of the barrel above the screw, which allowed straightforward gravity feeds to be utilised. The sides of the feed opening in modern machines are usually made to blend tangentially into the bore of the barrel and, provided the distance between the base of the hopper and the screw is not too great, the opening can be a parallel bore of the same diameter as the screw. Preferably, the feed opening should be elongated in the direction of the screw axis so that its length is greater than the lead of the screw. In some machines, the taper of the hopper is continued through the barrel wall to meet the bore tangentially in its wall section.

Almost all modern extruders have liners in the barrel, and this liner is extended into the feed section. Thus the feed opening pierces the liner, and the openings in the liner and in the barrel wall are carefully machined to coincide as accurately as possible, and may be polished and chromium-plated to minimise sticking and simplify cleaning.

Special types of feed openings are sometimes recommended for certain materials, in particular for nylon, which is extremely hard at ambient temperature and could cause excessive loads to be thrown on the screw if particles became jammed between the barrel wall and the outside diameter of the screw[1],[2].

In Section 3.2 a brief outline of the mechanism of feeding in a single-screw machine was given, from which it may be appreciated that the material which is fed to the screws must not be allowed to rise in temperature too rapidy. In fact it is the granules at the rear end of the screw which develop much of the forward thrust to generate pressure at the die; therefore, if the granular material were allowed to become soft and sticky it would adhere to the screw and forward motion would become difficult, or even impossible. In most single-screw machines as mentioned above, the feed section is water-jacketed to maintain a low temperature, and thus to obviate such feeding difficulties. Since the feed section is continuous with the rest of the machine barrel, which is heated, there is always a tendency for its temperature to rise if left to attain its equilibrium state.

In multi-screw machines, the screws are able to feed regardless of the length of the 'granular zone', and the necessity of maintaining a low temperature in the feed section, therefore, is not so great. In fact, it is often preferable to heat this section in view of the large proportion of heat which must be contributed by conduction on machines of this type.

6.3.2 MAIN SECTION

The main barrel section of the extruder is both a pressure vessel and a heating chamber, since the material is subjected to a gradually increasing pressure and suffers a considerable temperature rise during its passage through this part of the machine. The barrel, therefore, must be able to withstand pressures of up to 140 MN/m^2 (20 000 lbf/in^2) in some cases and be capable of transferring heat both to and from the material passing through. In addition to these pressure and thermal considerations, the problem of mechanical wear[3] by the screw and chemical action by the various plastic materials and/or their decomposition products must also be taken into account.

In order to withstand the high internal pressures resulting from the process, the barrel section has the form of a thick-walled tube, which is often a casting in the case of large machines and may be machined from solid rolled steel for small machines. The heavy wall section is also necessary from the point of view of rigidity and, furthermore, ensures that unsupported dies mounted on the end of the machine do not produce an appreciable deflection. In some cases in older machines the main barrel is made in sections which are bolted together by means of flanges or special clamp rings, and this sectional method of construction enables screws in differing length/diameter ratios to be used. It is, however, always necessary to keep the size of the clamp rings or flanges as small as possible, in order to minimise heat losses by radiation. Generally, the flanges are made circular so that heat losses are symmetrical, and so that band heaters may be easily fitted around them if required.

The majority of modern single-screw machines are equipped with cooling facilities on the barrel, and the simple cylindrical form of the main section has to be considerably modified to cater for these arrangements. Cored cooling channels are used in the case of machines which have cast barrel sections, and some form of fabricated jacket system may be used on machines utilising rolled steel barrel sections. A further method consists of cutting a helix round the outside of the barrel in which a copper tube is laid in intimate contact with the steel; the cooling medium is then passed through the tube. In one construction the barrel heaters themselves provide the cooling fins and channels for cooling fluid. The cooling channels in the barrel may also be used to convey heating fluids when the barrel heating is achieved by oil or some similar heat transfer fluid. It is usual to control the heating and cooling channels in zones by means of flow control valves so that a temperature gradient may be established along the machine barrel.

The thick-walled barrel of the extruder surrounds a hardened steel liner inside which the screw rotates. It is important that good contact be achieved between the liner and the barrel to ensure that the overall thermal conductivity of the total barrel wall is not impaired. The inside surface of the liner is usually highly polished to allow for the easy flow of material but in some special-purpose machines the liner has a bore which is rifled or grooved longitudinally. These grooves increase the friction between the material and the barrel wall, and thus enable a higher forward thrust to be obtained on materials which are difficult to feed.

The liner in a two-screw machine is formed in two sections, which together form a figure eight, and the inner surface in this type of machine is invariably smooth although it may be stepped.

Some thermoplastics materials—in particular polyvinylidene chloride and to a lesser extent polyvinyl chloride—emit gaseous decomposition products at high temperature which tend to corrode carbon steels. Thus, ordinary hardened steels are particularly susceptible to such corrosion. In addition to making liners hard to resist wear, therefore, it is also necessary to ensure that they are made of corrosion-resistant material.

A large number of extruders are now therefore fitted with seamless steel tube liners having an inner centrifugally cast lining of Xaloy, which is a special corrosion-resistant alloy developed in the U.S.A. This material is a low melting point hard material which can be machined with carbide cutting tools to a very smooth finish.

Two grades of Xaloy are available, one abrasion-resistant and the other, Xaloy 306, highly corrosion-resistant.

One of the advantages of liners made in the manner described above is that the structure of the hard inner surface is quite uniform and no change in properties results as the surface is worn away. Xaloy materials are, however, rather brittle and it is advisable to use a fairly thick barrel wall to minimise the risk of movement and consequent liner breakage. Liners made from Xaloy are usually at least 1.6 mm (one sixteenth of an inch) thick as the difficulties of production for thinner layers are very great.

Other liner alloys such as Hastalloy, Duranickel and Stoody are also used in special purpose machines, but these materials, although highly resistant to corrosion, are usually softer and more expensive.

Apart from the special liner alloys discussed above extruder barrel liners are also made from nitriding steels, case-hardened alloy steels and chilled cast iron.

6.3.3   DIE ADAPTOR DESIGN

The die adaptor of an extruder is that part of the machine which lies between the barrel and the die profile, and some parts of this assembly are sometimes referred to as the head, or die body. Because of the relatively complex nature of many dies, the die and die body (head) are often to all intents and purposes the same thing, and the whole assembly attached to the end of the barrel is frequently referred to merely as the die. A simple die arrangement is shown attached to the diagram of an extruder in *Figure 2.1* and can properly be said to include the breaker plate and screen pack assembly. It is essential that the die adaptor should be easily detachable to facilitate the removal and replacement of the gauze screens and that it should allow the rigid and leakproof attachment of a variety of dies.

There are many different methods of attaching the adaptor to the barrel of the machine, ranging from straightforward bolting, as implied but not shown in *Figure 2.1*, to various types of collet nut, bayonet attachments, split clamps, and even hinged fittings[4]. This latter method gives one of the easiest means of access to the screens and can be so designed as to allow rapid changes of the die, which is always hot and difficult to handle. It is essential in adaptor design, as in barrel flange design, to avoid irregular-shaped masses of unheated metal which could cause local heat loss and consequent cold spots at what is perhaps the most critical point in the temperature gradient. The majority of the attachment methods given above are therefore designed as symmetrical arrangements or, if this is not possible, provision is made to apply a specially shaped heater to any large protuberance which might conceivably dissipate too much heat.

The outside contours of the die adaptors are shaped in such a manner that heaters can easily be applied to them, and provision is usually made to fit a thermocouple or some controlling device so that the temperature of this section may be maintained at any required level. If the die is very small the temperature of the die orifice is determined and controlled by the setting of the instrument which controls the temperature of the die body. In other cases, however, the die orifice itself would be provided with one or more separate heater bands and thermocouples by means of which the die temperature could be controlled.

The inside contours of the adaptor and die body are suitably shaped to promote easy flow of the molten plastics material towards the die orifice. There are certain L/D ratios which should be aimed at, in order to avoid excessively convergent angles in the flow path, and these ratios are discussed in Chapter 8 and elsewhere[5]. As a general rule, the higher the melt viscosity of the material being extruded, the more acute must the flow-path angles be. Thus, very obtuse angles could be permitted for nylon, as in *Figure 6.1a*, but in the case of an unplasticised PVC material the included angle of flow path would have to be much more acute, as shown in *Figure 6.1b*. In the latter case it might be considered an advantage to fit a protrusion to the breaker plate, as shown dotted, in order to avoid a large mass of hot material.

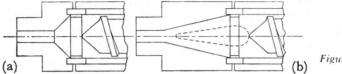

(a)　　　　　　　　　　　　　　　　　　　　　(b)　　*Figure 6.1*

The inner surface of the body is frequently chrome- or nickel-plated when used with corrosive materials.

The breaker plate is also part of the die assembly, and consists of a round plate which fits partly into the end of the barrel seating on the end of the liner and partly into the end of the die body, which it often serves to locate. The sum of the depths of the recesses in the barrel and the die body, into which the breaker plate fits, is arranged to be slightly less than the thickness of the breaker plate itself, so that a small clearance remains between the flanges when the whole assembly is bolted together. This clearance ensures that the breaker plate is rigidly clamped between the end liner faces and the die, and that no molten plastic material can leak between the faces and stagnate.

These features are discussed in greater detail in Chapter 8, which also contains further information on breaker plate design.

Screens of various mesh and wire sizes are usually used with the breaker plate and are placed on the screw side of the plate so that the pressure of the material maintains them in position. There is usually a space between the end of the screw and the breaker plate to accommodate the screens and, sometimes, the plate is recessed to receive them.

Usually a coarse mesh screen is placed against the breaker plate to serve as a support and to prevent bursting of the smaller mesh screens which are placed on top to give the required fineness. The screens normally used are made of stainless steel wire. A commonly used screen pack might consist of 1 x 100 mesh screen backed up by one of 60 mesh and one of 30.

Theoretically, provided the screw is of correct design, the screen pack should only be necessary to filter the material to ensure that no foreign matter or unmelted particles of material pass through to the die. However, in practice it is seldom possible to design a screw for each die arrangement, so that the required mixing in the metering section of the screw cannot be achieved at the pressures obtained when operating without a screen pack or some other form of pressure control. The screen device, therefore, in addition to being a filter, acts as a form of pressure mechanism to enable one screw to be used with a range of die orifices and control can be obtained by increasing or decreasing the fineness of the gauzes of which the screen is composed. This system is not a good one however as the fine screens tend to clog rapidly. A better method is to use a normal screen for filtering backed up by some form of controllable valve in the adaptor as discussed in Chapter 3.

The need for frequent changes of screen, when working with scrap material for example, has resulted in the development of several systems which enable this change to be made without dismantling the die, and several such systems have been described[6,7].

The screen changing systems referred to above and described in detail in refs. 6 and 7 are generally of two types. The first consists of a hydraulically operated slide on which two breaker plates complete with screen packs are mounted; operation of the slide moves a clean pack into position and at the same time moves the dirty screen to the outside of the extruder head where the gauzes are accessible for replacement.

The second type is more correctly called a screen 'cleaner' rather than a screen changer and consists of a separate unit containing one or more screen packs permanently in position with manually operated valving means to direct the melt flow from one to the other as required, and a backflow flushing system whereby a small part of the melt flow can be passed in reverse direction through the dirty screen and out via a purging valve. It is stated that by this means it is possible to run the extruders without changing the screen packs and that changeover from one screen to the other and the flushing of the dirty screens can be effected without interruption of the process.

In recent years a third screen system has been developed which is substantially different from the general methods described above. In basic essentials this 'Autoscreen' system consists of a band of steel gauze of appropriate mesh size which may be passed either continuously at low speed, or intermittently through the melt passage leading to the die. Thus the screen is in effect changed continually without interruption to the extrusion process, and cannot become blocked with foreign matter or unmelted polymer particles, which are brought out of the machine with the used portion of the filter band[8].

With all screen changing systems it is necessary to preheat the new screening medium to processing temperature, to avoid disturbing the melt flow, and to install a pressure indicating means upstream of the screen pack to determine when a clean screen is required.

6.4 Heating and heat control

One of the major requirements of an extrusion machine for thermoplastics is that it should be able to raise the temperature of the material passing through it at a controllable rate. Furthermore it is essential that the rate of heating should be adjustable to suit the speed of working and the thermal characteristics of the material being extruded. It is obvious therefore that an extruder must be equipped with a heating system which can be accurately controlled over a wide temperature range.

There are three methods of heating extruders and, in order of popularity for plastics materials, they are:
1. Electric heating
2. Fluid heating
3. Steam heating

6.4.1 ELECTRIC HEATING

This method is the latest historically, but it has such obvious advantages from the point of view of cleanliness, temperature range obtainable, ease of maintenance, economy of equipment, and overall efficiency, that it has now practically displaced both other forms of heating. The usual method of application is to fit resistance band heaters or, more recently, induction heaters around the outside of the machine barrel and die and to control these heaters by instruments which sense the temperature by means of thermocouples or resistance bulbs embedded in the barrel wall. It is normal to control the heaters on the extruder barrel in zones, so that the various parts may be held at different temperatures, in order to maintain a temperature gradient from feed to die. In this way the gradient set along the barrel can approximate to the gradient in the plastic material, and the rate of heat input can be maintained fairly constant along the whole of the barrel.

6.4.1.1 *Resistance heating*

The original type of band heater to be used on extruder barrels consisted of nichrome of other resistance wire insulated with mica strips and encased in flexible sheet steel covers. Such heaters, because of their compactness, flexibility and low price, are still popular. They are, however, fragile and given to the development of short circuits and other faults, at difficult times and will only withstand a loading of some 25-30 kW/m² (15-20 W/in²).

The efficiency and life of this type of heater depends on obtaining good tight contact with the metal of the barrel at all points. Failure to observe this requirement, besides resulting in irregular heating of the barrel, will also cause the band to overheat and destroy itself.

Band heaters with ceramics insulation, being rather more robust compared than the mica-insulated type, give a much better service life and also allow a higher loading. Unfortunately ceramic heaters are not flexible so that they are usually supplied in halves to bolt together around the barrel. Ceramic-insulated heaters in knuckle joint construction overcome this limitation.

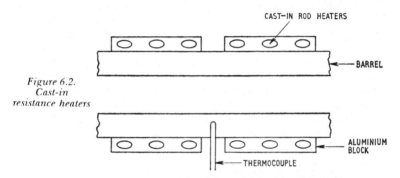

Figure 6.2.
Cast-in
resistance heaters

Another development is the 'cast-in' heater. In this type the insulated heating elements are cast into semi-circular or flat aluminium blocks which are accurately machined to mate with the surface to be heated. The heat from the element diffuses readily through the highly conductive aluminium and gives more even heating. This type of heater, which is illustrated in *Figure 6.2* is very robust and gives good service life.

6.4.1.2   Induction heating

This form of electrical heating does not function by the direct resistance of a wire element to the electrical current and the subsequent dissipation of this heat by conduction, but is generated by passing the normal a.c. mains frequency through an induction coil and thus setting up an alternating magnetic flux in an iron or steel associated conductor. Heat is generated from the resistance offered to the eddy currents set up by the magnetic flux.

In using induction heating for an extruder, the barrel becomes part of the system, i.e. it is the associated conductor and is heated directly by its resistance to the induced current and not by conduction.

The frequency of the a.c. supply controls the depth of heat generation but in an inverse ratio so that the greater the frequency the less the depth of heat generation. For normal frequencies of 50-60 Hz a depth of about 25 mm (1 in) is obtained, which is not far short of the thickness of many barrels. The amount of heat conduction required is therefore very small.

Other advantages of this system are that there is no heat lag and simple on/off instruments will give accurate temperature control. The possibility of hot spots or cold spots occurring is removed and power consumption is reduced although the power factor is below that of resistance heaters. The lower power consumption is due to the greater efficiency of the system as a whole and in particular to the considerably lower heat loss. With this form of heating, as it is not necessary for the induction coil to be in contact with the barrel, it is possible to design the cooling system to operate directly on the barrel surface. In this way a temperature control system, accurate over an exceptionally wide range, is possible. A disadvantage of induction heating is its relatively high cost. A schematic arrangement of a typical heater of this type is shown in *Figure 6.3.*

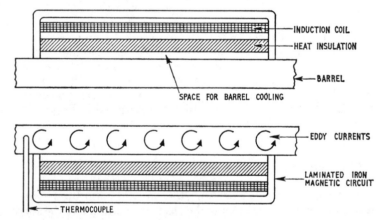

Figure 6.3. Schematic arrangement showing an induction heater in section

There are several methods of control available for electrical heating systems, of which the commonest are:

1. Manual control
2. 'On-off' controller
3. Proportional controller

6.4.1.3 Manual control

This method utilises a rheostat or a Variac-type transformer, and a temperature measuring device which is able to indicate the temperature of the zone under control. The machine operator manually adjusts the rheostat or the transformer until the rate of heat supply is balanced by the heat absorbed by the material passing through the machine and a steady temperature is recorded on the indicating instrument. The same method of adjustment is applied to each of the control zones on the machine, and in that way a continuous and constant heat input is obtained. A variation of this system uses Sunvic-type on-off regulators in place of the rheostat or Variac and one indicating instrument only, which can be switched to each zone as required.

Difficulties arise with these systems if the extrusion process is interrupted or if conditions inside the extruder become discontinuous. For, regardless of changes in the requirements of the material inside the machine, the system continues to apply the same amount of heat energy. It is for reasons such as this that systems 2 and 3 have been developed, which attempt to modify the heat input in accordance with the temperature of the barrel.

6.4.1.4 'On-off' controller

This method depends upon the operation of a control instrument which switches the heaters on or off as the indicated temperature, sensed by thermocouple or temperature-sensitive element, rises above or falls below the set temperature. In this way it is possible to maintain the temperature of the extruder barrel within certain limits which include the value set on the control instrument. Thus if conditions inside the machine vary, and the heat requirements change, this form of control automatically adjusts the heat input to suit the varying conditions. However, it is impossible to avoid a certain degree of hunting with this control since the thermal inertia of the extruder barrel and heater band inevitably causes a temperature override after the power supply has cut in or out. Furthermore, the fact that the whole heater load for a particular zone comes on or off as the switching device operates makes it extremely difficult to achieve a very fine control. On the other hand by using dual thermocouples connected in parallel, one as close as possible to the material and the other on the heater surface, a better control is obtained. The resultant of the two readings reduces the normal fluctuation to a steady average and enables the control system to operate from this figure.

6.4.1.5   *Proportional controller*

This method of operation depends upon the control instrument gradually throttling the power supply as the difference between the indicated temperature and the set temperature decreases. Thus, the energy supply to the extruder barrel is directly proportional to the difference between the control and the indicated temperature and when both coincide the power supply is completely cut off. This method, which works extremely well for small variations in energy requirements, is probably the most accurate for resistance heating systems as it is certainly the most expensive. A modified method uses a third relay to operate the automatic cooling system.

6.4.1.6   *Cascade temperature control*

A recent development in the field of heat control is the cascade system, which is linked to a direct temperature measurement of the material being extruded. This system uses a thermocouple which protrudes into the melt at the die adaptor position and the signal from this unit is compared with a preset instruction, variations being corrected on each heat zone in a cascade fashion. Screw speed variation can also be included in the same controls.

6.4.2   FLUID HEATING

The fluid most commonly used for heating extruders is oil, which may be heated by any suitable means. There are a number of heat transfer fluids available which have special properties inasmuch as they do not scale or clog supply pipes or can be used over wide temperature ranges but it is estimated that over 80 per cent of fluid heated extruders operate with oil. The heating system consists of a heater, a circulating pump, a surge tank, a filter, various valves, and the heat transfer channels in the extruder barrel. The construction of the extruder barrel is discussed in Section 6.3 wherein different types of heating and cooling channel arrangements are described. It is possible to use a fluid system to heat and cool, merely by passing fluid of the correct temperature through the heating channels, whereas it is normally necessary to incorporate an additional water or air cooling system on electrically heated machines.

The heat transfer unit, which contains the heater pump, surge tank and temperature controller, is quite separate from the extruder, to which it is connected by pipes and electric cables. Temperature control is obtained either by varying the rate of flow of a constant-temperature fluid or by varying the temperature of a constant supply of fluid. There is a considerable thermal lag in the operation of such a control system although when once conditions are established it has great stability.

6.4.3 STEAM HEATING

This method of heating is probably the earliest historically, but is now the least popular for thermoplastics. The high specific heat and latent heat of vaporisation of steam and water make them excellent heat transfer media but the disadvantages attached to their use far outweigh the advantages. Very few factories are equipped with steam-raising plant of sufficiently high pressure to give temperatures up to 300°C, as required by some thermoplastics, which in any case would be extremely inefficient; this, in addition to a tendency to corrosion, makes steam the least used method of extruder heating.

6.5 Cooling

On modern extruders with their high material shear, long barrels and high speeds, the cooling systems are frequently as important—if not more so—as the heating systems. Many of the highly critical techniques which are now in widespread use such as sheet extrusion, tubular film and rigid pipe production become very troublesome in operation if the stock temperature is allowed to become too high. At high throughput rates therefore the problem is how to prevent this temperature rise and short of reducing the output the only means is an effective system of cooling.

6.5.1 SCREW COOLING

In Section 3.2.3 reference was made to the mechanism of granular motion in the feed end of the single-screw machine and it was shown that this movement is dependent on the difference between the coefficients of friction of the granules on the screw and on the barrel wall. Since it is necessary that the frictional force of the barrel on the granules should be greater than that of the screw on the granules it is preferable that the screw should be maintained at a lower temperature than the barrel. However, if the screw is allowed to reach an equilibrium state, it eventually reaches a temperature higher than the barrel, because of the heat generated by shear in the material. In a single-screw machine, therefore, it is usual practice to cool the screw—or at least the feed zone thereof—in order to maintain it at a temperature below that of the barrel wall.

The screw is bored from the feed end, as shown in *Figure 6.4,* and a feed tube is inserted in the screw bore. Water is then either fed to the feed tube to return between the tube and the bore or is fed to the bore and returns by the tube. Certain advantages have been claimed for both methods but the differences in quality or power requirements between the two systems are relatively small. Water is fed to the screw either by means of a rotary gland which has both inlet and outlet pipe connections or it is fed and removed in the manner indicated in *Figure 6.4.* The cooling channel is not always taken right to the end of the screw and on smear-headed screws usually stops at the commencement of the torpedo.

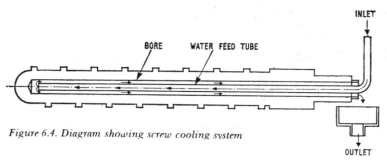

Figure 6.4. Diagram showing screw cooling system

Circulating oil has also been used for screw cooling in an attempt to overcome the corrosion and furring-up problems associated with water systems. This method also gives the added advantage of allowing the temperature of the screw coolant to be controlled at the level which gives the best output whilst still maintaining stability. This temperature is often just a few degrees below the barrel temperature. Direct cooling of the screw by cold water often results in a severe reduction of output but the controlled temperature method avoid this.

It is not usually considered necessary to cool the screws of a multi-screw machine since the feeding of materials in this type of equipment is independent of differential coefficients of friction. However, on at least one make of two-screw machine provision is made for screw cooling in order to obtain a certain measure of control over the temperature of the extruded material.

An interesting sealed cooling system has also been described whereby the hollow screw bore is filled to one-third of capacity with water at less than atmospheric pressure which is then impelled to the screw tip by an internal thread form. This cools the tip and the water vapour so produced travels to the rear where it condenses and heats the feed section of the screw. An adjustable plug is provided to vary the heat exchange rate[9].

6.5.2 BARREL COOLING

The possibility of the screw generating sufficient heat to raise the material to extrusion temperature is discussed at length in Section 3.8. However, under certain circumstances it is possible for the screw to generate more heat in the material than is required for the satisfactory working of the process and it becomes necessary to remove this excess heat by some means. The material characteristics which lead to overheating are low specific heat and high viscosity at working temperatures. Since it is virtually impossible to design one screw to work a material satisfactorily from a thermal point of view over a wide speed range a compromise screw is used and cooling must often be applied to the barrel in the upper speed range. For materials such as unplasticised PVC, barrel cooling is often imperative, because of its high viscosity and its susceptibility to degradation if held above certain temperatures for undue periods of time.

A description of various methods of applying cooling media to the extruder barrel is given in Section 6.3.2, where the actual construction of the main section is discussed. Ideally the cooling channels in the cylinder wall are controlled in zones in the same manner as the heaters and the rate of flow of cooling medium can be controlled by valves. In this way the required gradient can be maintained along the barrel by removing excess heat at different rates in each zone. It is also possible to control the cooling zones electrically in the same way that the heating zones are operated. Thus, a certain control temperature can be set on a barrel zone and if this temperature is exceeded the control instrument is arranged to supply cooling medium to the zone through a solenoid-operated valve. In this way the removal of excess heat can be catered for automatically and a close degree of control can be achieved.

Barrel-cooling facilities are not always provided on multi-screw machines since the possibility of generating excessive heat in the screws is remote, for reasons outlined in Chapter 4. The Trudex machine was equipped in this way, however, because of the construction of the screws which do in fact generate heat at one position.

The method and desirability of cooling the feed section of single-screw machines were outlined in Section 6.3.1 but it should be pointed out that automatic temperature control is not usually provided on this section. The requirement here is simply to maintain the section at a low temperature to prevent material melting in the feed aperture and to assist feeding; a constant flow of cold water is, therefore, sufficient.

A variety of cooling systems are used for extruder barrels, the most popular method being the use of a number of fans, one to each controlled zone, located on the underside of the barrel and arranged to blow air at room temperature over the barrel and heater surfaces. Automatic control is provided to operate the fans individually when the indicated temperature exceeds the set point on the control instrument. With cast-in heaters the casting can be finned on the outside to provide the maximum surface area. On heavy duty extruders, however, with their high generation of frictioned heat, an air coolant system may not be sufficiently effective and other methods are required. There are certain problems with water coolant systems in that they are more abrupt and less stable in their effect but because of the higher heat dissipation factor of water, heavy duty machines are frequently equipped in this way.

One such system uses copper tubing embedded in the heater or even in the barrel wall for water circulation and by using a double-threaded steel sleeve shrunk on to the barrel, cooling fluid can be made to circulate around the barrel in both directions, ensuring an even temperature over the zone. Cooling may be effected by passing the fluid through a heat exchanger in response to the thermocouple signal.

A patented vapour cooling system is also employed on one make of extruder. In this system the latent heat of the vapour which circulates around the barrel is extracted by an automatically controlled water cooling system which surrounds a condensing chamber situated outside of the barrel. This method is said to give a very smoothly operating and effective control.

6.5.3   THRUST RACE COOLING

In common with many types of bearing arrangements on rotating machinery, provision is made to flush the thrust race and journal bearings of the extruder screw, not only to lubricate these points but also to act as a coolant. Because of the high pressures generated in the barrel and the consequent heavy load on the thrust race this component usually operates under fairly rigorous conditions. It is, therefore, normal practice on both single- and multi-screw extruders to provide an oil-circulating pump, which maintains a constant flow of oil through the bearings to dissipate any heat generated by the high loading conditions. The necessity for cooling the thrust assembly is more imperative for multi-screw than for single-screw machines because of the bearing limitations of the former.

6.6   Hopper design

6.6.1   PLAIN FEEDER

The commonest form of feed hopper used on single-screw machines is the conical type of standard design. The material feeds into the screw merely by its own weight and consequently, this type of hopper is frequently referred to as a gravity feed.

The hopper may be fitted with a lid to keep the material dust- and moisture-free and with a window in order that the material level may be seen. A shut-off gate is usually provided to the base of the hopper, by means of which the supply of material to the screw may be stopped, or regulated, and in some machines a drying or pre-heating unit is fitted which blows hot air through the material in order to remove moisture before feeding to the screw or in some cases to raise the temperature of the feed material.

This type of hopper feed has two main disadvantages: firstly, the height of the material in the hopper is continually varying and this produces slight changes in the pressure conditions in the feed end of the screw which can consequently affect the overall machine performance; secondly, even with the hopper itself and the feed port to the screw correctly designed there is always danger of material bridging and thus producing an erratic feed, or even a complete cessation of feed. The first difficulty is inherent in this type of hopper and can only be overcome if the material is continually maintained to a fixed level.

Several systems are available for automatically maintaining the hopper level and include venturi feed systems and secondary hoppers. The second difficulty can be overcome by fitting an agitator or stirrer in the hopper, driven by a small electric motor or by fitting a vibratory attachment. The inverted cone is the most common shape for gravity-feed hoppers and the degree of taper should

be as steep as possible consistent with capacity; tapers between 20°
and 45° are commonly used.

6.6.2 CONTINUOUS FEEDER

One type of continuous feed is achieved by means of vibrating
feeder, which is sometimes adopted to overcome the difficulty mentioned in Section 6.6.1 in connection with the changing head of
material in a gravity-feed hopper.

In one form of this equipment the material feeds from the base
of a separately mounted hopper on to a longitudinal platform
which is supported on flat springs at an angle with the vertical.
The armature of an electromagnet is attached to the inside of the
platform, and when a pulsed d.c. or single-phase rectified a.c. current is supplied to the coil of the magnet the platform vibrates in a
plane perpendicular to the plane of the flat springs. As a result of
this vibration the material moves along the platform, falls over the
end into the feed aperture, and thence into the extruder screw.

By varying the amplitude of the vibrations the rate of feed can be
adjusted so that a constant head of material is maintained in the
feed opening. In this way constant feed conditons can be maintained regardless of the height of material in the primary hopper.

Another type of continuous feed is used as an integral part of a
drying system. Materials such as cellulose acetate, which must be
well dried before feeding, are often fed by a vibrator to an endless
belt on which they are conveyed under radiant heaters and finally
tipped directly into the extruder hopper. This method also allows
the head of material in the hopper to be maintained at a constant
level if the speed of the feed belt is synchronised with the speed of
the extruder.

For the feeding of difficult materials such as chopped film
scrap, chopped film scrap mixed with a proportion of virgin granules, some PVC dry blends, etc. and sometimes in order to increase output or to obtain an acceptable output from materials of
low bulk density, special force feed hoppers have been designed.
Sometimes known as 'crammer' feeders, these hoppers are provided with a vertical screw with variable speed drive, to 'cram' the
materials into the extruder feed section, and a vibrator or agitator
to ensure the filling of the crammer screw.

6.6.3 SPECIAL FEED CONSIDERATIONS FOR MULTI-SCREW MACHINES

In Chapter 4 mention was made of the fact that it is not always advisable to fill the screw flights of a multi-screw machine completely because of the danger of overloading the drive. Consequently, it is usual on this type of equipment to fit a feed mechanism which can be adjusted to fill the screws to the desired amount and to maintain the feed at a constant rate. Some multi-screw machines are equipped with adjustable feed devices which are geared to the screw drive and, therefore, work at a rate which is directly proportional to the screw speed.

The Colombo machine is fitted with an oscillating type of feed which meters a certain amount of material from the hopper into the screws at each oscillation. The frequency of oscillation is determined by the speed of the machine screws but it is possible to adjust the amplitude of the oscillating member, and thus to vary the amount fed at each throw, to allow for materials of different bulk factors and to modify the extrusion conditions.

The Trudex and other machines utilise a screw type of feed mechanism. The hoppers of these machines are equipped with single screws which are geared to the main drive by variable speed boxes of the positive infinitely variable (P.I.V) type. The screw acts as a metering conveyor and delivers material to the main screws. In this system the speed ratio between the main screws and the feed screw is varied by means of the P.I.V. box to cater for the differences in the bulk factor of various materials.

6.6.4 VACUUM HOPPER

With the increasing use of powder having a small particle size and the techniques of dry blending for PVC, modified equipment has been developed to deal with the problems of entrapped air and moisture which often cause poor surface finish. The vacuum hopper was introduced in 1959 by N.R.M. with the object of removing vapours etc. from the mix prior to its entering the extruder. The hopper is of conventional design but is sealed and placed under vacuum. The material is fed into the hopper from a secondary storage source, in such a manner that the vacuum is not destroyed, and a seal is also required in the extruder between the cylinder barrel and the screw bearing. An air-operated heavy-duty feed valve regulates the material feed to the hopper and material level is maintained by manual control or by means of a level-sensing device which actuates the feed mechanism automatically. Results using vacuum hoppers have been reported as showing improvement in both surface finish properties and extrusion speeds,[10,11,12].

6.7 Thrust bearing features

6.7.1 SINGLE-SCREW MACHINES

The generation of pressure is fundamental to the extrusion process and is achieved by the rotating screw inside the machine barrel. The radial components of this pressure are catered for by the thick barrel walls and the forward component pushes the material through the die of the machine to produce the required shape. There is, however, an equal and opposite reaction to this forward thrust, which is exerted on the screw of the machine, and the magnitude of this force is often of the order of 1.1MN (250 000 lb) or more on a 150 mm (6 in) machine. Because of this large end-load on the screw, special arrangements have to be made to accommodate very high capacity thrust bearings at the rear of the screw. Since the general tendency is to operate machines at higher speeds, greater loads are thrown on to the thrust-race arrangement and the bearings themselves have to be capable of transmitting high loads at higher speeds.

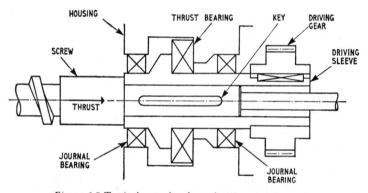

Figure 6.5 Typical extruder thrust-bearing arrangement

Figure 6.5 shows, diagrammatically, a typical bearing arrangement for a single-screw machine. The screw of the machine is usually keyed or splined at the rear end and fits into a driving sleeve in the bearing housing. The driving sleeve rotates in ball races and is also able to bear against a thrust race which can transmit the thrust from the screw directly to the bearing housing.

Figure 6.6 shows two arrangements using tapered roller bearings in which the larger bearings (steep angle) take the thrust and are backed up by the smaller bearing to take the radial load.

Early machines were normally fitted with an ordinary ball-type thrust race but as the capacity of machines has developed the size of thrust race has correspondingly increased and the ball type is no longer adequate. Some manufacturers use a single-taper roller race in the thrust assembly in an attempt to accommodate both radial and axial loads in the same bearing. Generally speaking, however, this type of bearing does not have a sufficiently high axial capacity, and some type of separate thrust bearing is usually fitted. Very high capacity thrust bearings are available which utilise cylindrical or barrel-shaped rollers in place of balls, thus allowing a much greater load-carrying surface.

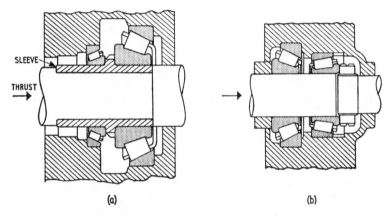

Figure 6.6. Diagrams showing two typical thrust bearing assemblies using tapered rollers

Thrust bearings are designed to operate under specific load conditions and their life expectancy can be estimated with a fair degree of accuracy under these conditions. Overloading of thrust bearings will reduce their service life out of all proportion to the excess load and it is suggested that the service life of a bearing varies inversely as the cube of the load. It is important therefore that thrust bearings should always be of a capacity well in excess of the anticipated maximum load.

Double and tandem bearings which are common on multi-screw machines have also been employed on single screws where it is expected that the machine will be operating under extreme conditions. A patented thrust bearing assembly using two equally dimensioned races in tandem has been described[13]. This arrangement uses a hydraulic system to equalise the thrust on the two bearings and at the same time to adjust the axial movement of the screw to a predetermined back-pressure.

As mentioned in Section 6.5.3, special arrangements are often made to cool the thrust race and journal races, by means of an oil-circulating system, in order to ensure the optimum working conditions for these parts. This necessitates taking adequate precautions against oil leaking into the barrel of the machine and thus contaminating the plastics stock. An efficient oil seal is fitted on the screw side of the thrust housing and the barrel-hopper section. This separation of the two parts of the machine reduces to a minimum the risk of contamination and also serves to prevent particles of plastics material entering the thrust housing, besides assisting the cooling of this member.

6.7.2   MULTI-SCREW MACHINES

The problem of accommodating heavy axial loads on a multi-screw machine is even more difficult than on the single screw. The largest diameter of thrust race which can be used on a multi-screw machine is limited by the small clearance between the screws. Therefore, since the capacity of a thrust bearing is usually determined by its diameter, normal bearing assemblies for multi-screw machine applications have a very definite capacity limit. The thrust-bearing positions on multi-screw machines are normally staggered so that one bearing lies in front of the other, thus permitting larger-diameter races to be used, and certain types are fitted with special thrust assemblies.

Since the screw positions on multi-screw machines impose such a definite limitation on bearing diameters, the thrust loads on

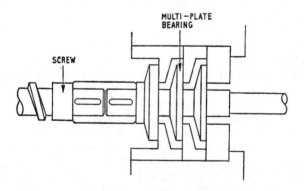

Figure 6.7. One type of thrust bearing arrangement for two-screw machines. (Courtesy R.H. Windsor Ltd)

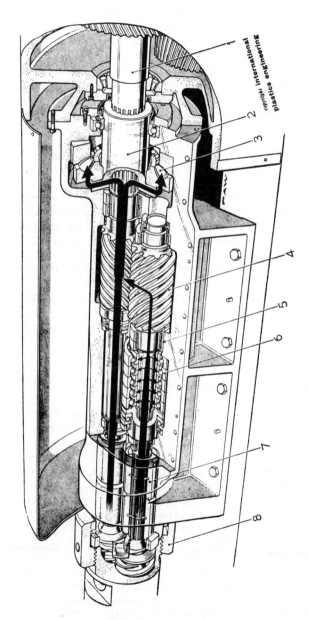

Figure 6.8. Heavy twin-screw thrust bearing system as used on the Mapre 160 mm machine. Int. Plast. Enging)

some makes of machine are taken up on multi-layer bearings which depend to a large extent on length for their capacity. Such an arrangement is shown diagrammatically in *Figure 6.7* from which it may be seen that the bearing utilises sliding surfaces, arranged in layers. This type of bearing gives rise to problems of alignment, machining and lubrication. The difficulties of obtaining an equal load distribution over each of the bearing surfaces will be readily apparent and, since there are two and sometimes three of these assemblies to each machine, the manufacturing cost is obviously high. Furthermore, it is imperative to provide some form of forced lubrication since the plain bearing surfaces must be continually lubricated. Other makes of multi-screw machines use force-lubricated conical-thrust assemblies to increase the bearing area.

Figure 6.8 shows an even more elaborate bearing assembly in which the thrust on one screw is first reduced by a series of roller and ball thrust bearings and these transferred to the shaft of the second screw which is extended and can therefore accommodate a larger thrust bearing.

In Section 4.3 a converging twin-screw arrangement is described which results in a greater distance between the screw spindles at the rear of the machine. Unfortunately, the angle at which the screws can be pitched is relatively small and no great increase in bearing sizes and capacity normally results from this arrangement.

6.8 Reduction gear and drive motor

The range of screw speeds lies between 10 and 200 rev/min for a single-screw machine and between 5 and 50 rev/min for a multi-screw machine. Therefore, since the majority of extruders are powered by electric motors with output speeds of 1000 rev/min or more it is necessary to build a high-ratio reduction gear into the machine drive. A common form of reduction gear for extruders up to 75 or 100 mm (3 or 4 in) in diameter is a worm and wheel of standard design, which is easily built into the machine. Ratios of the magnitude required are easily obtained in these units and the fact that they are standard pieces of equipment relieves the extruder manufacturer of the necessity of producing a relatively complex piece of gearing. Moreover, a standard worm reduction box is a sturdy self-contained unit which normally contains its own lubrication system and which can be expected to function for many years with the minimum of attention.

Machines of above 100 mm (4 in) screw diameter are normally fitted with drives which have to transmit 60 kW (80 hp) or more and a worm reduction unit designed to transmit such power to an output shaft turning at 10-100 rev/min tends to become very bulky. The tendency is, therefore, to use a multi-stage herring-bone gear reducer, which can be more conveniently arranged than the worm unit. Furthermore, the efficiency of herring-bone gearing is higher than that of worm gearing and this becomes an increasingly important consideration as the transmitted power increases.

The manufacturer of a multi-screw machine is inevitably committed to producing a certain amount of gearing in order to drive the screws in the required direction. Probably, for this reason, on this type of machine the reduction system, often consisting of a train of spur or helical gearing, is built into the machine as an integral part of the drive mechanism.

Not only is it necessary to fit a speed-reducing mechanism on an extruder; it is also essential that facilities should be available for varying the speed of the screw over a wide range without reducing too drastically its torque. Certain materials require slower speeds of processing and in others the higher outputs of high screw speeds are important.

There are three main methods of obtaining a stepless speed change and in each case the prime mover is an electric motor. The first method is purely electrical, or electronic, in operation and consists of a system whereby the motor output speed may be varied over the required range, as for example with an a.c. commutator motor or a Ward-Leonard set, of which there are many variations available. In each of these systems the motor is both the prime mover and the speed-changing device and in each case the power output of the motor is roughly proportional to the output speed. Thus, at low screw speeds the available power from the motor is low, which for certain hard materials is a decided disadvantage unless a motor larger than is necessary at high speeds is fitted to cater for this low-speed limitation.

The second method of obtaining a screw-speed variation consists of a mechanical friction drive driven by a constant speed electric motor. There are several types of friction drive suitable for smaller extruders, but there are very few capable of transmitting more than 11 kW (15 hp), which in general limits their application to machines of 63 mm (2½ in) and below. Large P.I.V. belt drives and Beier gear units are made, capable of transmitting more than 100 kW (150 hp), but complications arise when using these units at low speeds because of the high torque transmitted. The great advantage of friction-type drives is that a constant power electric motor can be used as the prime mover, and, provided the drive is capable of transmitting the power at low speeds, constant power is available over the whole speed range. In practice, very few variable-speed devices are capable of transmitting full power at low speeds and special shear pin or slipping clutch mechanisms have to be installed in the drive to protect the screw and other parts of the transmission against excessively high torque at these low speeds. Combinations of these methods are also commonly used. Constant speed motors fitted with gearboxes or change gears give six or more screw speeds but no adjustment whilst the machine is running. Two- or three-speed gearboxes have been fitted to variable-speed motors to increase the power available at low speeds and this probably represents the best compromise arrangement.

The third method of obtaining controlled power input and screw-speed variation is by hydraulic means wherein a constant-speed electric motor is used to drive a piston-type, positive-displacement, variable-volume, hydraulic pump which provides the hydraulic power required to drive a fixed-displacement piston-type hydraulic motor. The hydraulic motor can be coupled directly to the extruder screw—thus dispensing with the reduction gear—or for convenience of design it may be linked thereto by a one to one gear set. It is possible by this means to control the speed of the hydraulic motor—and thus of the screw—from a maximum of the order of 200 rev/min down to creep speeds, without steps, by varying the delivery of oil from the pump to the hydraulic motor.

Drives of this type have constant torque characteristics and the maximum operating torque level can be controlled by means of a relief valve which can also serve as a safety valve to prevent overloading of the extruder screw and gearing.

6.8.1 FINAL DRIVE

Vee belts, direct coupling and sometimes chains are all used to provide the final drive between the motor and reduction gear on commercial extruders. Chains and direct couplings provide the most positive drive but do not insulate the machine from drive vibration and fail to provide a safeguard in the event of machine overloading.

Belt drives on the other hand, although not so positive as the direct coupling or chain drive method, do provide insulation against shocks and can sometimes slip in the event of a very severe overload. Belt drives also are more flexible as to drive arrangements, are easier to install. and are not so critical as to alignment.

Finally it is usual to include an ammeter in the drive motor circuit to indicate the power being used or to give warning of an overload. An ammeter is suitable for the latter purpose but only gives an approximation of the power. A wattmeter is needed to give a true indication of the power taken by the motor but very few machines are yet so equipped. An equally important item of drive installation which is sometimes omitted or forgotten is the tachometer to indicate the screw speed. Most machines are now equipped with good tachometers but older machines, many of which are still in use, are not so provided.

6.9 Machine base

The machine base of an extruder is the structure on which the barrel, reduction gear and other parts are supported. In some cases it may also provide a housing for the control instruments and a cover for the drive motor.

Although the base is in effect just a hollow metal structure, it is important that it should be designed correctly because should distortion occur under load the whole functioning of the extruder could be adversely affected due to misalignment.

In some cases the base is fabricated from steel plate and in others it is cast. If the former method is used, the structure must be of ample thickness and suitably webbed and buttressed to minimise vibration and 'spring'. Generally, however, a well matured casting is to be preferred because a base made in this way is usually very rigid and free from 'spring'.

If any of the control instruments are built on to the machine base they must be very well insulated from vibration and should also be so placed that they are protected from spanners etc. which may be dropped during die adjustment or change. It is in fact always preferable to house all the control instruments in a separate cabinet well removed from the extruder itself.

Housing the drive motor in the machine base is commonly practised and provides a very neat and streamlined arrangement. It is quite unnecessary to do this, however, and such a construction frequently creates trouble in design later due to the difficulty of motor cooling or because of the need to fit a motor which is too large to be housed in the space available. An external drive motor is generally the most practical and flexible arrangement.

6.10 Computer control of the extrusion process

This chapter has, so far, dealt with the constructional features of extruders and has considered in some detail the standard and orthodox methods of control of the many process variables which arise. It will have been noticed that there are many such variables, all of which have an effect on the quality of the finished product and on the rate of output. The temperatures at the various positions on the barrel, for example, must be controlled to a nicety, which including the die adaptor section could involve perhaps six control points: the temperature of the die itself must also be accurately controlled and in the case of a wide sheet die this could include a similar number of control positions. The temperature of the screw or of the screw cooling medium, the temperature of the water cooling jacket surrounding the material feed section and even the temperature of the feed material itself are all important factors. Additionally the screw speed must be controlled, the melt pressure at the die head must at least be measured as an important parameter and the speed of take-off must be adjusted to control product dimensions.

These controllable variables, which are all interrelated, are, in an orthodox extruder, controlled manually by the exercise of operator skill and experience. This method, although reasonably successful, has obvious limitations in that adequately skilled oper- ators are rare and becoming rarer and that such skills vary greatly from operator to operator and from day to day.'

These limitations would be avoided and the extrusion process would become much more certain and economically viable if the many process parameters could be controlled non-manually from some central point. With the advent and rapid growth of computer technology during recent years such a control system has now become possible.

There are now two principal systems for computer control of the extrusion process. One is the direct digital system wherein the computer takes over the full control function of the extruder and thus the instruments typically found in temperature-control cabinets and other control instrumentation are eliminated and the computer becomes the reception point of all the readings and acts upon these.

The second system is a supervisory control in which standard controlling instruments are employed to perform the direct control function of the process whilst the computer operates in a supervisory capacity, continually updating the set points on the controllers according to the conditions obtaining at that time. This system has the advantage that should the computer fail it can be easily switched out of the circuit and local manual control reapplied.

The most sophisticated computer control systems, which at the time of writing have already been installed in several extrusion plants, provide continuous closed loop control of temperatures in ten or more positions, material feed rate and screw and take-off speeds, Measurements of the process parameters are taken by normal analog controls and fed to a digital computer, via an electronic interface, which then transmits any necessary adjustments back to the machine.

Such systems are already in use or in planning, for the extrusion of pipe, sheet and films and the degree of sophistication can be varied to suit requirements. A closed-loop system can be employed for the control of certain temperatures only, for example, with an open-loop system for the screw speed, leaving the take-off speed and certain die temperatures to be adjusted manually. However, with the further development of computer and extrusion technologies it seems certain that many important extrusion plants will soon go over to completely computerised closed-loop control systems for all the process parameters and it is not difficult to visualise a time in the future when die thickness and concentricity adjustments are also similarly controlled.

6.11   Ultrasonic control

Another interesting development which has been proposed is the control of the extrudate and the extrusion process by the use of ultrasonic feedback. The dynamic viscosity of the melt is measured during the actual operation of the extruder by means of a small ultrasonic generator which measures the pulse speed and the absorption coefficient of the material as the ultrasonic waves pass through the melt. This information can then be fed into a control unit and used to adjust the process variables of the extruder. In tests carried out both temperature and pressure measurements were taken at the same point at the ultrasonic generator[14].

REFERENCES

1   N.V. ONDERZOEKINGSINSTITUUT RESEARCH, Br. Pat. 721413 (30 September 1952)

2   TOLL, K.G., 'Shape Extrusion Studies of Zytel Nylon Resin', *S.P.E. tech. Pap.*, **3** (1957); *S.P.E. Jl*, **13**, 17 (1957)

3   MADDOCK, B.H., 'The Effect of Wear on Delivery Capacity of Extruder-Screws', *S.P.E. Jl*, **15**, 433 (1959)

4   FISHER, E.G., 'Trends in Extrusion Machinery', *Br. Plast. mould. Prod. Trader*, **26**, 297 (1953)

5   FISHER, E.G., in *Polythene*, Ed. Renfrew, A. and Morgan, P., Iliffe, London (1960), p. 843

6   SCHUTZ, F.C., 'Extruder Screening Devices', *S.P.E. J1*, **19**, 547 (1963)

7   VOIGHT, B.R., 'The Case of Screen Pack Changers', *Mod. Plast.*, **43**, 125 (1966)

8   Autoscreen Process Developments Ltd, London, maker's literature

9   SCHLOEMANN, A.G., Düsseldorf, Br. Pat. 1 173 738 (12 November 1968)

10   MUNDY, W.M., 'Vacuum Feed Extrusion', *S.P.E. J1*, **15**, 887 (1959)

11   FLATHERS, N.T. *et al.*, 'Vacuum Hopper Extrusion', *Mod. Plast.*, **37**, 105 (1960)

12   FLATHERS, N.T. *et al.* 'Advances in Dry Blend Extrusion', *Mod. Plast.*, **38**, 210 (1961)

13   ANON., 'Amigo 2 in. Hydraulic Drive Extruder'. *Int. Plast. Engng*, **2**, 92 (1962)

14   PETRIB, M., *Sixth Symposium of Polymer Processing, March 1972*

7

Materials for Extrusion

7.1 General

The number of plastics materials available for processing by extrusion is continually growing, as is the range of finished or semi-finished products which may be fabricated by this method. Extrusion is probably the most versatile process available to the plastics industry, and it is also a high-output-rate operation. The plastics material suppliers, therefore, are generally prepared to devote considerable time and money to the development of special extrusion grades in any new material they may produce and to the investigation of new extrusion applications. The specialist extrusion firms have also shown great ingenuity in the development of extrusion processes, often in combination with some other manipulative method, and in the production of new or unusual items or in the establishment of new uses. Extruded garden fencing and trellis, hollow skirting boards with closable recesses for electric wiring, extruded plastic netting as described elsewhere, and extruded radio cabinets are just a very few of the many unusual items which have been noted.

Broadly speaking there are now 16 common materials or groups of materials which are commercially processed by extrusion techniques:

1 Acrylic resins (polymethyl methacrylate)
2 ABS copolymers (acrylonitrile-butadiene-styrene copolymers)
3 Casein plastics
4 Cellulosic materials (cellulose acetate; cellulose acetate butyrate, cellulose propionate, etc)
5 Foamed plastics (foamed polystyrene; foamed polyvinyl chloride; foamed polyolefins etc)

6 Polyacetals
7 Polyamides
8 Polycarbonates
9 Polyolefins, including the low, medium and high density polyethylenes, polypropylene, the polybutenes and copolymers
10 Polystyrenes (including modified polystyrenes)
11 Polyvinylidene-chloride-based materials
12 Vinyl plastics (both homo- and copolymers in plasticised or flexible form)
13 Vinyl plastics (both homo- and copolymers in rigid unplasticised form)
14 Fluorocarbon resins
15 Urethane elastomers
16 Thermosetting materials (including reinforced plastics)

It will be convenient to discuss each of the above materials or groups of materials separately, commenting briefly on any special points which arise in their processing. This, where possible, will be followed in each case by brief details, in general terms, of the range of products into which the particular material is normally extruded.

Casein extrusion has been adequately described in a monograph devoted to this material[1] and the extrusion of thermosetting resins forms the subject of Chapter 11 of the present monograph. No further discussion on these plastics is, therefore, necessary here.

Apart from the materials listed above, other polymers such as chlorinated polyethers, polyallomers, ionomer resins, phenoxy resins, polysulphones, polyphenylene oxides and a number of others, have also been successfully extruded and possess interesting properties in the finished form. The commercial importance of these polymers in the extruded form is not yet sufficiently developed to warrant individual attention in this monograph.

7.2 Extrusion compounds

Materials for extrusion are often supplied by the manufacturers in the form of specially formulated compounds. These materials which are usually granular in form, may contain heat and light stabilisers, lubricants, pigments, plasticisers, and other additives—in addition to the basic resin—which either improve their extrusion properties or impart the characteristics required for the particular end product or both.

The compounding operation may also be carried out by the processor using special equipment and in a few cases it can be a continuous process in the extrusion machine.

Depending on the requirements of the basic resin, the compounding process consists broadly of carefully dispersing the various ingredients by means of suitable mixing equipment and then fluxing the mix so that it becomes an homogeneous mass. This mass may then be sheeted on heated rolls and finally cut into uniform granules using conventional size-reduction equipment. The fluxing can be carried out either in an internal mixer of the Banbury type, on open heated rolls, or even in extrusion machines of one form or another as described in Chapter 5. Attention must be given to the following requirements in the extrusion compound.

7.2.1 FLOW

The material must flow into a homogeneous melt without hard, unmixed, or imperfectly mixed particles and, particularly in the case of vinyl materials, must show no undue tendency to adhere to the heated metal parts of the extrusion machine. These factors are influenced by good compounding, the correct choice of plasticiser and the correct degree of internal lubrication, all of which have been adequately discussed elsewhere[2],[3].

7.2.2 LUBRICATION

The correct lubricant, used in the right proportions, is important inasmuch as if the compound is overlubricated correct shear effects will not take place in the screw. The degree of lubrication required depends partly on the type of extrusion machine used.

7.2.3 STABILITY

The compound must be adequately stabilised against heat and light so that degradation does not occur in the extrusion machine or during the service life of the finished product. The stabilisers, plasticisers and lubricants themselves must of course also be equally stable against these effects, as must the pigments or other colours.

7.2.4 GRANULE OR PARTICLE SHAPE AND SIZE

Extrusion materials are available in a number of different forms as described in detail in Chapter 10. They may be supplied as free-running powders, in random-cut chips, or in regular cubes, cylinders or spheres[4]. There is some controversy regarding the best form and size for such compounds, and much depends on the type of thermoplastic and on the characteristics of the extrusion machine being used. Generally, however, it is agreed that plasticised vinyls, polyethylene, cellulosic materials, and nylon give the best results in regular 3 mm or 2.25 mm (1/8 in or 3/32 in) cubes, cylinders, or spheres; whereas unplasticised PVC and polystyrene can be, with advantage, somewhat smaller.

7.2.5 MOISTURE CONTENT

The percentage of moisture in a material for extrusion is a very important factor[5]. If this figure exceeds certain low limits then the product will suffer from many obscure faults which are difficult to diagnose. In the worse cases obvious steam bubbles will actually form and burst in the extrusion as it leaves the die but in less seriously damp materials, the moisture may show itself as lines of minute bubbles to give an unevenly matt surface; and in transparent extrusions will cause cloudiness among other defects. Occasionally, an erratic surging effect may also be caused by a high moisture content.

Some thermoplastics, such as the cellulosics and nylon, are very hygroscopic whilst others—the acrylics—are slightly so. These materials must always be dried immediately before extrusion on non-vented machines. Other materials such as the vinyls and polyethylenes do not normally absorb moisture and so may usually be extruded as received from the supplier. Polystyrene, also, is considered to be non-hygroscopic but does, nevertheless, have a tendency to take up moisture on the surface of the granules; for best results, therefore, it should also be dried. A factor which is often overlooked in this matter of drying is the question of the compounding ingredients which may themselves be moisture-absorbent although the basic resin is not. Thus, the filler, etc., in a vinyl compound may give trouble as may the pigments incorporated into a coloured polyethylene. Figures for maximum moisture content for satisfactory extrusion are issued by the raw material manufacturers and should be carefully adhered to. Drying may be carried out in various ways and there are a number of firms specialising in drying equipment for thermoplastics. The simplest, very popular, method uses a fan-ventilated controlled-temperature oven with a number of shallow trays. Drying time, depending on the material, varies from two to four hours at about 80-85 °C for normal low temperature thermoplastics whilst nylon requires a temperature of about 110-120 °C for the same period, preferably under vacuum.

The use of vented extruders with facilities for the removal of moisture and other volatiles by vacuum extraction via either the barrel or the screw, as described in an earlier chapter (or, in special cases, the use of vacuum hoppers) is a very desirable precaution which will often avoid the need for an elaborate predrying operation.

7.2.6 TESTING OF EXTRUSION COMPOUNDS

The melt flow properties of thermoplastics extrusion compounds vary greatly, not only between the different grades of a particular material, but also within one supposedly uniform grade, and often from bag to bag of a particular batch of one grade. This variation increases the difficulty of extrusion and can also affect the physical properties of the final product. It is of value therefore to be able to assess rapidly on the shop floor the melt flow properties of fresh batches of material prior to use so that the necessary adjustments can be made to the production equipment. Instruments known as plastometers are used for such rheological testing but these machines function by determining the pressure required to force a melt at a known temperature through a standard-size orifice[6], and difficulty is often encountered in correlating this information to the performance of a screw extruder of production size. To overcome this difficulty a number of interesting test instruments have been developed, and of these perhaps the most widely known is the Brabender Plastograph/Plasti-Corder. This instrument is a recording torque rheometer by means of which polymers and polymer compounds can be analysed in terms of viscosity and flow and sample-sized batches of material can be subjected to test conditions which approximate to those encountered in commercial processing. A wide range of interchangeable heads is available on this equipment including mixing systems of several types and extruders with a range of interchangeable screws of differing lengths and flight configurations.

In considering this important problem of the flow properties of materials for extrusion it seems obvious that an extruder operating adiabatically, as described in Chapter 3, and adequately instrumented, could form the basis of a very informative test instrument for this purpose and several workers have in consequence developed equipment on these lines.

One of the most interesting of these test instruments was the equipment developed in the early 1960s by Fraser and Glass Ltd, called the 'Evaluator'[7,8]: this machine was in fact a small high speed extruder equipped with a micrometer adjustment to a cone head/cone screw tip combination to control pressure, and an adjustable feed mechanism to control the rate of input of material. Screw speed was accurately adjustable, and indicated, up to a maximum of 900 rev/min and the drive motor current was also indicated. External heating by conduction was supplied to the head only and was switched off automatically at a preset level, depending on the material under examination, and the extruder then operated completely adiabatically. The method of test with this equipment was to feed the extrusion material to the screw at a uniform and known rate and then to manipulate the micrometer head setting until a satisfactory extrudate was obtained. The characteristics of the material were then related to the output rate, the head setting, screw speed and drive motor current[9].

7.3 Notes on the extrusion of various thermoplastics

The extrusion conditions to be observed in actual processing vary considerably according to a number of factors. Extrusion machines differ, for example, in the way in which they work on different materials and basic resins differ from maker to maker. The formulation of the compound, the type of granule, and the class of product will also have their effects on the conditions of extrusion.

The notes in this chapter therefore are given as a guide only.

7.3.1 ACRYLIC RESINS

These thermoplastics are noted particularly for their clarity, and extrusions in them are often required primarily for this feature. On the other hand, acrylic resins are notoriously difficult to extrude in such a way that this clarity is preserved. The main difficulty is in the tendency for the polymer to depolymerise under the combined effects of friction and heat, giving rise to bubbles and cloudiness. The extrusion conditions and operation must, therefore, aim at holding the frictional effects to a minimum whilst still supplying sufficient heat to allow correct extrusion. The heat must be supplied as far as possible by conduction, so that a deep-cut screw is often used and the screw speed is normally kept low. In order to give the required controlled heating over an extended period without steep temperature gradients, extruders with long barrels (*L/D* ratio 20 : 1 minimum) have been found satisfactory[10]. Both metering-type screws and short-transition-zone screws have been successfully used, the most important feature being the channel depth, which must be considered in relation to the compression ratio. Work by Garber and Cassidy on the effect of metering screws in the processing of acrylic resins has led to a better understanding of the effectiveness of this combination[11].

Mention must also be made of the importance of vented extruders in the processing of acrylics. Vented machines allow the removal of volatiles from the melt, in this case the products of depolymerisation, in addition to traces of moisture, and thus lead to an extrudate of improved quality at increased output rates. If vented extruders are not used, then the acrylic resin should be dried to a moisture content of less than 0.25% immediately prior to extrusion and the use of breaker plates and screen packs avoided in order to reduce the risk of depolymerisation.

7.3.1.1 Applications

Acrylic resins are normally extruded into rod stock and tube; diffuser trough fittings for fluorescent lighting and into corrugated and flat sheet.

7.3.2 ABS COPOLYMERS

These materials are terpolymers with a uniform molecular structure and therefore differ from other modified polystyrenes which may be blends of polymers. For best results predrying of the material is recommended, or the use of a vented extruder. Extrusion temperature has been found to be somewhat critical within the range 175-230 °C; variations in processing temperatures outside this band have a noticeable effect on surface finish. Metering-type screws give a satisfactory extrudate and smear-head screws can be recommended to give adequate back-pressure if no breaker plate or filter is used. A point to be noted with the extrusion of these materials is that they are subject to rapid freeze, therefore the calibration device or take-off equipment must be located in close proximity to the die face.

7.3.2.1 Applications

ABS materials are extruded into two major forms: flat sheet for further manipulation by thermoforming for travel cases, refrigerator liners and structural components: and pipe for conveyance of natural gas, water and chemicals.

7.3.3 CELLULOSICS

The only materials in this group which are of commercial importance for extrusion in the U.K. are cellulose acetate and cellulose acetate butyrate. Both materials extrude without undue trouble on normal equipment with screws whose L/D ratios are 20 : 1 or higher, and both smear-headed and short-metering-section types have been used successfully. The most troublesome characteristic of these materials, particularly the acetate, is their tendency to absorb moisture from the atmosphere. Adequate predrying to a moisture content of less than 0.25% or the use of a vented extruder[12], or both, is therefore essential to a satisfactory extrusion operation.

7.3.3.1 Applications

Cellulosic materials are extruded into a wide range of forms, the most important of these being tubes and profiles for decorative purposes and for covering handrails and the like. Pipelines for use in oil fields and also for the conveyance of town and natural gas, tubing for irrigation, sheet for vacuum and other forming, and film and tubular film products for packaging are also produced.

7.3.4 FOAMED PLASTICS

The extrusion of foamed thermoplastics is at the present time a comparatively small-volume operation, but its importance is rapidly growing, and the products so produced would seem to have considerable future potential.

In the brief description which follows, it will be convenient to divide the extrudable foamed thermoplastics into two groups according to the method of expansion.

7.3.4.1 Chemical blowing

The first group of materials consists of those in which a chemical blowing agent is incorporated in the thermoplastics compound or is dry-mixed with the compounded granules prior to extrusion. During the extrusion process the blowing agent decomposes under the effect of heat and liberates a gas under pressure — usually nitrogen — into the molten material. Foamed polyolefins and foamed rigid vinyls are commonly produced in this way and most other thermoplastics can be treated in a similar manner to produce interesting lightweight materials.

The principal requirement for the successful extrusion of the above foamed materials is accurate temperature control of the melt to prevent premature decomposition of the blowing agent. The thermoplastic melt/gas mixture so produced remains in solution, owing to pressure in the machine, until it emerges from the die when the gas comes out of solution and rapidly expands to cause frothing of the extrudate. By accurately controlling the temperature of the melt it is possible to initiate the gassing effect at different stages in the melt flow path so that varying amounts of residual gas are entrapped giving a degree of control over foam density.

This foaming process has been further developed, as will be discussed in a later chapter, whereby the potentially 'frothing' and somewhat uncontrollable extrudate passes directly from the extrusion die into a water-cooled or other calibrating system wherein the product is chilled and sized to give a foamed lightweight interior with a polished and entire outer surface of accurate dimensions[13],[14].

7.3.4.2 Physical blowing

The second group of thermoplastics materials of interest to the plastics extrusion industry includes those in which the polymer or compound granules contain a 'physical' blowing agent such as a volatile hydrocarbon which evaporates when heated to expand the heat-softened granules into hollow particles or beads and finally diffuses away through the resulting membranous granule wall. The expanded granules or beads so formed can be fused together subsequently in closed mould systems — in practice usually by the action of steam or hot air — to produce inexpensive lightweight moulded forms which are of considerable value and much used for the protection of delicate articles in the packaging industry. The most important material to be foamed or expanded in this way is, at the time of writing, polystyrene and the fusing of the expanded beads, and in fact the whole process, including the introduction of the blowing agent, can be carried out in an extruder for the production of expanded continuous profiles, expanded film and sheet and composites which include an expanded polystyrene layer.

Further details on the technology of foam extrusion are given in a following chapter dealing with complete extrusion processes.

7.3.4.3 *Applications*

Foamed plastics based on the polyolefins are used as wire and cable insulation and foamed vinyl materials are used both as insulators and as fillers in multi-core cable construction. High density cellular vinyl and polystyrene extrusions are finding growing use as replacements for wood in many applications including profiles, window frames, doors etc. in the building industry and as weathering strips, gaskets, etc. in automobile and aeronautical applications. 'Physically' blown expanded polystyrene is extruded into film and sheet for packaging and for decorative paper applications, and the production of parisons for blow moulding has been considered[21].

7.3.5 POLYACETALS

These are addition polymers of formaldehyde. These materials have high melt viscosity which is not appreciably altered by temperature variations, and have been successfully extruded on metering-type screw of L/D ratio 20 : 1 and above. It is important that both the extruder and the die used for processing these materials should be free of dead spots where the material could stagnate leading to discoloration and depolymerisation. Detailed studies of the effects of hold-up times at various temperatures have been reported[5] and theoretical work on the melting and freezing rates of polyacetal materials and their effect on extrusion conditions have been studied by Richardson[16].

7.3.5.1 *Applications*

Polyacetals can be extruded into sheet, rod, tubes or used in wire-coating applications. The rod stock is often used for later machining to produce automobile and electronic components for replacement of metal parts, thus utilising the good wearing properties, fatigue endurance, dimensional stability and chemical resistance of these materials.

7.3.6 NYLON-TYPE MATERIALS

There are now many grades of polyamidic and similar materials available to the processor, but all are, to a greater or lesser degree, characterised by their high processing temperatures, their rapid transition from solid to fluid within a narrow temperature range, their fluidity at melt, and their tendency to absorb moisture. The extrusion equipment and its operation must, therefore, be arranged to cope with these characteristics. The screw design and breaker plate restriction must be adequate to build up back-pressure with the low viscosity melt and the machine itself must be of ample capacity to maintain hydrostatic pressure in the die system. A minimum screw length of 20D is recommended with short transition and shallow channel depth. Heater capacity must be sufficient to cope with the high melting point and low shear rate. Pfluger[17] has considered in detail the effect of screw geometry with the various polyamides and this, together with the study by Toll[18] has led to a greater understanding of the extrusion technology concerning these materials.

The polyamides are hygroscopic and the granules are usually supplied vacuum-dried in sealed containers. Once these containers have been opened it is important that the contents be used without undue delay. In the event of a quantity of the material being left exposed to the atmosphere for some time, or in the reclamation of scrap, the material must be carefully re-dried before use.

An interesting development in the extrusion of nylon 6 and some other nylons is the production of high-molecular-weight finished strips, films, profiles, etc., directly from the raw materials by the continuous polymerisation of lactams in a specially developed twin-screw compounding extruder using an alkaline reaction.

The process is rather complicated in that the low viscosity liquid reaction mixture must be conveyed by the extruder against the viscous melt of the polymer as it is formed, a complex temperature ᵽrofile must be maintained on the barrel of the extruder which must also contain provision for the removal of volatiles and unreacted materials, and the degree of polymerisation must be closely controlled[23].

7.3.6.1 Applications

Polyamides and similar materials, in the extruded form, are characterised by their high tensile strength and resistance to abrasion. On the other hand, the materials tend to absorb moisture, are expensive and difficult to reclaim for re-use.

Applications of extruded polyamides include monofilament of various types; blown, water-quenched or chill-cast film for specialised packaging applications, rod stock for machining into tough abrasion-resistant components such as gears, cams and bearings, etc., tube and pipe for hydraulic oil lines and for the conveyance of oils and petrol.

7.3.7 POLYCARBONATES

These materials are polymers derived from bisphenol A. The main properties of the polycarbonates of interest in relation to their extrusion performance are high processing temperatures (250-300 °C), very high melt viscosity and sensitivity to moisture. The extrusion equipment, therefore, must be of rugged construction with ample drive power and capable of working at high temperature. Because of the moisture sensitivity of the material the machine should preferably be of the vented type and equipped with a hopper predryer. In general, material care and drying precautions as used with the nylons should be observed. The polycarbonates are not critical as regards screw design and several standard types, as used with the higher density polyolefins, have been used with success. Adequate back-pressure is developed without screens or breaker plates and screw cooling is not recommended.

7.3.7.1 Applications

The most important advantages of the polycarbonate resins are high rigidity and toughness, high impact strength combined with good clarity, heat resistance, dimensional stability and reasonable electrical properties. Current applications in which the combination of these properties in the extruded form has proved valuable include tubing, rod stock, film and sheet, for specialised uses in the electronics, aeronautical and automobile industries and in vandal-proof glazing.

7.3.8 POLYETHYLENES

The generic term polyethylene covers an important group of the polyolefin thermoplastics materials which are available in a wide range of densities and melt flow properties suited to a variety of applications and processing techniques.

Generally speaking the polyethylenes are also among the easiest materials to extrude and the good results which can be obtained on any reasonable machine are reflected in the vast quantity of extruded polyethylene products currently available. In order to obtain optimum output and quality with the higher density materials, however, a long-metering-type screw (L/D ratio not less than 20 : 1) is recommended, with a compression ratio between 2.5 and 3.5 : 1 dependent upon the final extrudate form. For outdoor applications where weathering resistance is important, polyethylene is usually compounded with an antioxidant together with 2% of hard carbon black to give protection against the effects of ultraviolet light. Owing to the tendency towards moisture absorption by the carbon-black component of this mixture, preheating prior to extrusion is recommended.

7.3.8.1 *Applications*

The polyethylenes are probably the most widely used of all plastics materials and a high proportion of their applications result from the extrusion process. The most important usage for low density polyethylene is in the packaging industry, where it is extruded into tubular and flat film in both standard and shrinkable varieties for wrapping an immense variety of products and, in the case of high-density, high-molecular-weight materials, into tissue 'paper' film and bags for the retail food trades. Polyethylene is also extrusion-laminated to paper, metal foils, cellophane and other substrates to give composite packaging media where the properties of each material are combined. Extruded polyethylenes are also widely used as low-dielectric-loss material for wire and cable insulation, and cross-linked grades of polyethylene have also been developed for this purpose. Extruded pipe and tubing from the various grades of polyethylene are also widely used. Mention must also be made of the blow moulding of high and low density polyethylene bottles for packaging applications, now an important branch of the industry, as blow moulding, which is described elsewhere in this book, is an extension of the extrusion process.

7.3.9 POLYPROPYLENE

Polypropylene, like the polyethylenes, is also a member of the polyolefin family of thermoplastics materials, but in this case the molecular structure can be tailored to give a desired range of properties. The processing of polypropylene is very similar to that of the polyethylenes and the same general recommendations are applicable. Polypropylene, however, develops less frictional heat in the extruder screw and consequently more external heat must generally be supplied by the barrel heaters.

7.3.9.1 *Applications*

Polypropylene has higher temperature resistance and toughness than the normal grades of polyethylene and many of its applications make use of these properties. As with the polyethylenes, polypropylene is extruded into films for packaging—usually by the chill roll casting process—which when oriented give low cost products of high strength and clarity, and which are rapidly gaining favour as replacements for Cellophane, in cigarette carton wrapping for example. Other important, recently developed applications for extruded polypropylene films are in the manufacture of 'tape-yarn'—split oriented film of approximately 3 mm width—which is used as a replacement for jute in the manufacture of sacks and carpet backing and in fibrillated material whereby the above-mentioned oriented tape-yarn is longitudinally disrupted to produce a staple fibre of use in carpet pile construction, etc. Another important development in which the unique strength properties of oriented extruded polypropylene are utilised is in the manufacture of strapping tapes which now are rapidly replacing steel tapes for the tying of cartons and boxes.

In addition to the above, extruded polypropylene also finds limited but important application in the form of monofilaments for cords and ropes, sheet for vacuum forming, wire and cable insulation and protection, and pipes and tubes.

Further information on some of the above important developments is given in a later chapter.

7.3.10 POLYSTYRENES

The polystyrenes cover a wide range of homopolymers, copolymers and blends of polymers which have been developed to improve the impact strength of the brittle basic resin. These materials give little trouble in extrusion on modern equipment with long screws, and generally the long metering type is preferred[19]. Special fluted torpedo designs (Dulmage) have also been recommended and are effective.

7.3.10.1 *Applications*

Normal polystyrene is not greatly used at the present time for ordinary tube and profile extrusion because of its pronounced brittleness. The high-impact and rubber-modified varieties are widely used, however, but show some limitation in that a clear high impact variety is not yet available. Normal grades of polystyrene are used in limited quantities for the extrusion of lamp troughs for fluorescent lighting, where they form a cheaper substitute for the extruded acrylic component previously mentioned, and for the production of oriented packaging film and brush bristles. The major extrusion usage of the high impact polystyrenes is in the production of flat sheet for thermoforming into numerous items including trays and containers for packaging, refrigerator body and door liners.

7.3.11 POLYVINYLIDENE CHLORIDE MATERIALS

Polyvinylidene chloride extrusion materials are not generally available in the U.K. These thermoplastics which enjoy some popularity in the U.S.A. for special uses, are believed to consist of unplasticised copolymers of vinylidene chloride with small proportions of other monomers such as vinyl chloride which are included to improve processability. Polyvinylidene chloride materials are generally available only as free-running powders and their extrusion presents certain special problems. The worst of these difficulties concerns the low heat stability of the basic polymer, particularly when in contact with steel at a high temperature.

In order to overcome this difficulty, the screw is usually quite short and deep to reduce the shear action, and terminates in a long unflighted cone. The form of this cone is followed quite accurately by the cone of the extrusion head and all parts which come into contact with the heated material are constructed of special non-ferrous alloys of which Hastalloy, Stellite and Duranickel are examples.

7.3.11.1 *Applications*

Polyvinylidene chloride materials, are extruded into water, petrol- and solvent-resistant tubing which finds use in the chemical industries and in hydraulic control systems, etc. It is also extruded into monofilaments, into films for special packaging and for inter-layers to composite laminates, for which application it possesses excellent, in some respects unique, barrier properties.

7.3.12 PLASTICISED PVC MATERIALS

Plasticised PVC is a compounded or fluxed mixture of the polymer, plasticiser, heat and light stabilisers, and lubricants. It may also contain pigments and fillers. The ease of extrusion depends to a large extent on the formulation and on the adequacy of mixing and fluxing during the compounding operation.

Provided these operations are correctly carried out, all plasticised PVC compounds.extrude well on normal equipment with either medium or long screws.

7.3.12.1 *Applications*

The range of applications for plasticised PVC extrusions is so vast that it is impossible to do justice to the subject in this book. Soft PVC extrusions in one form or another are used in almost every industry and are available as wire coverings, tubes, pipes, covering to products other than wire, profiles in an immense variety filaments of various types, and sheet and film for the furnishing and packaging industries. New applications are added daily.

7.3.12.2 *PVC dry blends*

No discussion on the extrusion of plasticised vinyl materials would be complete without at least a brief reference to 'dry blends'.

In the normal compounding process referred to in Section 7.2 the plasticiser which forms part of the PVC compound formulation is incorporated during the compounding and the final composition is fully fluxed to give a completely gelled complex. With certain PVC resins whose particle formation is so adapted that it can absorb plasticiser very rapidly this full gelling can be dispensed with and a satisfactory compound produced in which the plasticiser has been absorbed with only a very small amount of fluxing. Such compounds—called 'dry blends—are formed simply by mixing the heated resin powder with hot plasticiser until the mix becomes dry and free running. The other ingredients, such as lubricants, stabilisers, pigments, etc., must be incorporated at the correct stage and it is essential that the final dry blend is adequately cooled before bagging.

The advantage claimed for dry blends over completely gelled compounds include greater economy, reduced heat history and higher rates of extrusion. The subject is, however, a highly controversial one, with ardent supporters of both techniques. The success of a dry blend extrusion process depends greatly on the resin properties, the temperature and method of mixing and the order in which the various ingredients are added, in addition to the design of the extruder and the extrusion technique employed.

So far as extrusion is concerned, the processing of dry blends calls for a slightly different treatment than for compounds prepared by the orthodox method. As only a small amount of gelling has taken place in the manufacture of the dry blend the extruder must make up for this and contribute the necessary extra heat, and work, to the compound. This usually means that the screw must be of adequate length, i.e. not less than 20*D*, and have a higher compression ratio than for normal PVC compounds. A compression ratio of 4 : 1 has been found suitable for working with dry blends.

One important problem which frequently occurs in the extrusion of dry blends is the tendency towards porosity and poor surface finish. This is normally due to the presence of moisture and other volatiles which result from the greater and more receptive exposed surface area of the powder material as against that of normal granules. Various methods of overcoming this difficulty, by the use of vented extruders, screws of special design, pre-drying in a heated oven, etc., have been suggested, but the most practical method appears to be the use of a sealed vacuum hopper as described in Section 6.6.4 and discussed elsewhere[20].

7.3.12.3 *PVC colour concentrates*

It is common practice nowadays to colour PVC compounds in the form of granules by means of specially formulated compounds known as 'master batches', with a high pigment concentration. To do this, the processor merely mixes by tumbling, etc., the correct proportion of concentrate to an unpigmented PVC compound and feeds this mixture to the extruder. In this way the carrying of stocks of compounds in a range of colours is avoided. Similar concentrates are available for the colouring of polyethylene, polystyrene and other resins.

7.3.13 RIGID PVC MATERIALS

Rigid PVC may consist of PVC resin compounded with a small percentage of plasticiser—such as 5% or 7½%—together with stabilisers and lubricants, etc., or it may have no plasticiser. It may also be formulated with 'straight' PVC resin or with copolymer resins based on PVC or with a blend of both.

Unplasticised PVC is a difficult material to extrude as it normally possesses poor flow properties and has a pronounced tendency to decompose at the high temperatures necessary for its processing. Such materials must, therefore, be well stabilised and internally lubricated, the equipment must be designed so that interruptions to flow are minimised and all pockets and steps where materials may stagnate are carefully avoided.

Because of the thermal instability of unplasticised PVC resins the temperature of the extruder barrel and die parts must be controlled within close limits. If the temperature is too low, excess shear will be developed in the high viscosity material, causing local overheating and degradation; on the other hand, if the temperature is too high the material will be degraded by excess conducted heat. Successful extrusion of the unplasticised PVC therefore, depends on the attainment and preservation of a nice temperature balance within these narrow limits.

The inclusion of small proportions of plasticisers and/or the use of copolymer resins reduces these tendencies and thus simplifies the extrusion, although considerable care must still be used. On the other hand, the materials are then of less value to industry because of their lower softening temperatures, lessened chemical resistance and low impact strength.

Unplasticised, high molecular weight, straight polyvinyl chloride in powdered form—i.e. without prefluxing and granulating—are being increasingly used for the direct extrusion of rigid PVC pipe and sheet using both single- and multi-screw machines. In this way the heat history of the mix is kept to a minimum and the risk of thermal degradation reduced. On the other hand, certain workers have felt that this method may result in the production of pipe and sheet with reduced, or non-uniform, physical properties owing to lack of control of the final gelled state of the material. Because of this, and also because of the difficulty of obtaining uniform feed of powdered materials to single-screw machines, many firms still prefer to extrude their UPVC products from a granular material. In this case a partially gelled compound in which the PVC resin is merely pressed together rather than completely fluxed is desirable in order to retain as far as possible the advantage of reduced heat history of the resin. Little if any published literature is available on the comparative properties of pipes and other products extruded from these different forms of raw material.

7.3.13.1 Applications

Rigid PVC extrusions can be divided broadly into two groups according to the resin type and compound formulations as outlined above. The first group consists of rigid vinyl compounds which are formulated from copolymer resins or resin blends, and may also include a small proportion of plasticiser. These materials give good surface finish and the extrusions are widely used for the production of tubes and profiles, etc., in a range of colours, mainly for decorative purposes. Their softening temperatures are not high and their chemical resistance is reduced—for a rigid vinyl material—but they are cheap and satisfactory if correctly applied.

The second group consists of high molecular weight straight PVC resins without plasticiser. Extrusions in these materials are available as pipes and tubes for the chemical industry, domestic cold water supply, waste pipes, soil pipes and rainwater goods; flat and corrugated sheet for building applications and for fabrication into chemical-resistant plant; film for foodstuff packaging and miscellaneous profiles for special applications.

Special rubber-modified compounds in which the PVC resin is

blended with a small proportion of a special nitrile or other rubber are available for use in applications where extra-high impact resistance is desirable. These blended materials extrude readily although the precautions already outlined for normal unplasticised materials should be observed.

7.3.14 FLUOROCARBON RESINS

Although the fluorocarbon resins are of limited commercial importance at the present time, it is felt that their potential applications are sufficently interesting to warrent inclusion in this book. The four polymers which are discussed briefly below are perhaps the most important types, and are generally representative of the whole group, which is characterised by outstanding electrical properties, chemical inertness and heat resistance.

Polytetrafluorethylene (PTFE) was the first of these materials to be developed, and is still the most important of the fluorinated polymers. It is a true thermoplastic, but because of its high processing temperature (330°C) and difficult flow properties, it is not possible to obtain satisfactory extrusion in conventional screw machines. Ram or screw/ram systems of the type used for the extrusion of thermosets, and described in Chapter 11, have, therefore, been adapted for this process.

In one method of operation the polymer in powder or granular form is compacted into discontinuous preforms which are then sintered in a heated die to become solid rod or tube. Because of the very low production rate when using this method, modified screw machines have been developed for continuous extrusion[22]. In this technique the screw acts merely as a forwarding and compacting device to produce and feed a PTFE preform continuously to a heated elongated sintering die. The screw is designed with constant pitch and depth, and, therefore, has no compression ratio. The extrusion operation is, in fact, a complete reversal of conventional extrusion processes in that both the screw and barrel are cold and the material is heated after it has passed through the preform forming die to produce a homogeneous extrudate.

The other important fluorocarbon resins, i.e. polychlorotrifluoroethylene (PCTFE), tetrafluoroethylene/hexafluoropropylene copolymer (FEP) and polyvinylidene fluoride, although characterised by their high melt temperatures and high melt viscosities, have substantially normal flow properties and can, therefore, be processed on conventional equipment. Extruders for use with these materials must be designed to operate at high temperatures and pressures. Constant pitch, short compression zone screws of *L/D* ratio greater than 20 : 1 are usually recommended.

7.3.14.1 Applications.

Products extruded from fluorocarbon resins are mainly utilised for their excellent electrical properties, good chemical resistance and dimensional stability, all of which apply at elevated temperatures. Rod stocks are produced for the machining of gaskets, seals, tapes and small components for specialised applications; thin and thick walled tubing, wire and cable insulation, sheet and numerous small profiles are also available. Fluorocarbon resins are all very costly at present, and use of them is therefore reserved for essential applications in which their high cost is justified.

7.3.15 URETHANE MATERIALS

The urethane range of elastomers has become increasingly important over recent years, principally in injection moulding but also in extrusion. The extrusion of these materials presented some problems initially in that a surging effect was obtained despite rigid control of the machine parameters. From experimentation it is now proved that the best performance is obtained from a screw design having an approximately 5*D* feed section, a 5-8*D* transition section and a 7-10*D* metering section. The worst conditions were obtained when screws with rapid transition zones were used. The raw material should be kept as dry as possible to obtain acceptable surface finish, and careful temperature control is necessary to prevent overheating.

REFERENCES

1 COLLINS, J.H., *Casein Plastics and Allied Materials*, Plastics Monograph No. C5, 2nd edn, The Plastics Institute, London (1952)
2 BRETT, H.D., 'The Development of Plastics Extrusion, A Survey with Particular Reference to PVC', *Plastics, Lond.*, **18**, 406 (1953)
3 BRETT, H.D., 'PVC' Extrusion Compounds', *Plastics, Lond*, **21**, 15 (1956)
4 KENNAWAY, A., 'The Granulation of Polythene', *Br. Plast. mould Prod. Trader*, **28**, 18 (1955)
5 SIMONDS, H.R., WEITH, A.J. and SCHACK, W., *Extrusion of Plastics, Rubber and Metals*, Reinhold, New York (1952) p. 178
6 COLLINS, J.H. *Testing and Analysis of Plastics*, Plastics Monograph No. C2, The Plastics Institute, London (1954)

7   ANON., 'Adiabatic Extruder for Evaluation of Thermoplastics', *Int. Plast. Engng*, **1**, 239 (1961)

8   ANON., 'Extruder Evaluator for Colour Matching and PVC Stabilisation Tests', *Int. Plast. Engng*, **2**, 32 (1962)

9   SMALL, G.G., 'Mouldability of Rigid PVC Compounds) (Letter to the Editor), *Br. Plast.*, **35**, 324 (1962)

10 GRIFFITHS, L., 'Developments in the Processing and Application of Acrylics', *Plast. Inst. Trans.*, **27**, 193 (1959)

11 GARBER, J.F. and CASSIDY, R.T., 'Performance Characteristics of Metering Type Screws on Acrylic Materials', *S.P.E. tech. Pap.*, **8** (1962)

12 DEPCIK, H.W., 'Special Materials for the Extrusion', *Kunststoffe*, **52**, 284 (1962)

13 ANON, 'Structural Foam Goes Extruded', *Mod. Plast.*, **46**(1), 106 (1969)

14 UGINE KUHLMANN, Fr. Pat. 1498620 (19 September 1967)

15 BARRETT, G.F.C., 'Acetal Resins', *Plastics, Lond.*, **25**, 136 (1960)

16 RICHARDSON, P.N., 'Processing of Acetal Resins', *S.P.E. Jl*, **16**, 1324 (1960)

17 PFLUGER, R., 'Polyamides as Extrusion Materials', *Kunststoffe*, **52**, 273 (1962)

18 TOLL, K.G., 'Shape Extrusion Studies on Nylon Resin', *S.P.E. Jl*, **13**, 17 (1957)

19 GRIFF, A.L., 'Super-Impact Polystyrenes', *Plastics, Lond*, **27**, 57 (1962)

20 FLATHERS, N.T., JOHNSON, R.E., PALLAS, V.R. and MAYO SMITH, W., 'Vacuum Hopper Extrusion', *Mod. Plast.*, **37**, 105 (1960)

21 GOLDSBERRY, H.H. and FOX, A.J., 'Blow Moulding Expandable Polystyrene', *S.P.E. tech. Pap.*, **8** (1962)

22 ELLIOTT, E.M., 'Screw Extrusion of Granular Fluon Polytetrafluoroethylene', *Plastics, Lond.*, **23**, 8 (1958)

23 ILLING, G., 'Direct Extrusion of Nylon Products for Lactams', *Mod. Plast.*, **46**, 70 (1969)

8

Dies for Extrusion

8.1 General

In previous chapters of this work the extrusion machine has been discussed in some detail and the better known extrusion materials and their applications have been described. The quality of an extruded product, however, depends not only on the extruder and the material used, but also on the efficiency of the die and other pieces of equipment which are part of the process. This chapter deals with certain aspects of die design and gives a general outline of the form and construction of the types of die used in well-known extrusion processes.

8.2 Some aspects of die design theory

The function of an extrusion die is to form the molten material delivered by the screw into a required cross section. The die is, therefore, a channel whose profile changes from that of the extruder bore to an orifice which produces the required form. In order to predict the behaviour of a thermoplastics melt in such a channel it is necessary to know the viscosity of the melt over the required range of shear rates and temperatures, and to be able to relate this viscosity with the flow of the melt under pressure through the different sections of which the channel is composed. Then since the pressure drops are additive the total pressure drop and throughput through the whole assembly can be calculated.

8.2.1 NEWTONIAN AND NON-NEWTONIAN FLOW

The flow of a Newtonian fluid through a tube of circular cross sec-

Dies for Extrusion

tion is given by the Poiseuille equation as

$$v = \frac{dp}{dx} \cdot \frac{R^2 - r^2}{4\eta}$$ (8.1)

where R=outer radius and r=inner radius, by which it can be seen that the velocity is proportional ro r^2 and that the velocity profile across the tube is a paraboloid of revolution *(Figure 8.1).*

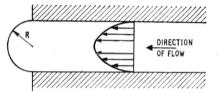

DIRECTION
OF FLOW

Figure 8.1. Velocity profile of Newtonian fluid in tube of circular cross section

It is well known, however, and has been discussed in previous chapters, that thermoplastics melts do not in general behave as Newtonian fluids, so that the formulae required to describe the flow of these materials in an extrusion die are considerably more complex than the simple Poiseuille expression given above. Several workers[1-4] have put forward interesting approximations in an attempt to simplify these expressions but despite this valuable work, the formulae are still somewhat difficult to handle in all but the simplest of circumstances.

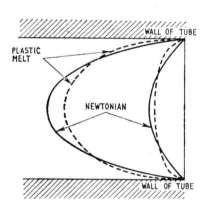

WALL OF TUBE

PLASTIC
MELT

NEWTONIAN

WALL OF TUBE

Figure 8.2. Velocity profiles for Newtonian fluids compared with those for polymer melts of two different viscosities

The velocity profiles of both Newtonian and non-Newtonian fluids are shown in *Figure 8.2,* and the considerable difference between these two types of flow can be readily seen. It will be noticed in particular that the velocity gradient at the tube walls where the rate of shear is highest is much steeper in the case of the non-Newtonian thermoplastics melts than with the true Newtonian fluid. From this it will be again apparent that the behaviour of thermoplastics melts departs from Newtonian flow as the rate of shear increases. Since the shear rates encountered in most thick-section extrusion dies working under normal condiitons are quite low, it follows that the relatively simple classical flow expression can still be used in many interesting die calculations. It is possible for example to compare the effects of changes in die path shape and the relative merits of increased land length as against reduced channel width as a means of restriction in particular cases.

Figure 8.3. Flow through a slit orifice

The discharge of a Newtonian fluid through a narrow rectangular channel as shown in *Figure 8.3* is given by

$$Q = \frac{ad^3}{k\eta} \frac{\mathrm{d}p}{\mathrm{d}x} \tag{8.2}$$

where Q = volumetric discharge, $\mathrm{d}p/\mathrm{d}x$ = pressure gradient, η = viscosity, a = width of slit, d = thickness of slit, and k is a parameter dependent on the ratio a/d. This expression gives a useful approximation to the flow of a plastics melt through a die of this form.

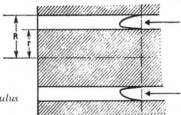

Figure 8.4. Flow in a circular annulus

If now the slit is considered to be wrapped into a cylinder around a radius r then a similar expression can be derived for tube dies, as illustrated in *Figure 8.4*.

$$Q = \frac{\pi}{k\eta}\left(\frac{R + r}{2}\right)(R - r)^3 \frac{\mathrm{d}p}{\mathrm{d}x}$$

or

$$Q = \frac{\pi}{k\eta} \cdot D_\mathrm{m} \cdot t^3 \frac{\mathrm{d}p}{\mathrm{d}x} \tag{8.3}$$

where D_m is the mean diameter of the tube and $t (= R - r)$ is the wall thickness. If Q is plotted against D_m for a series of wall thicknesses at a constant pressure gradient, it will be found that the plots, although curved at their origins, tend to become straight at D_m increases. This is clearly shown in *Figure 8.5*, which indicates an approximately linear relationship between output and tube diameter, whilst the effect of channel or land length, the x term in the formula, is linear also. The vertical distance between the plots, however, is proportional to t^3, indicating a cubic relationship between output and wall thickness.

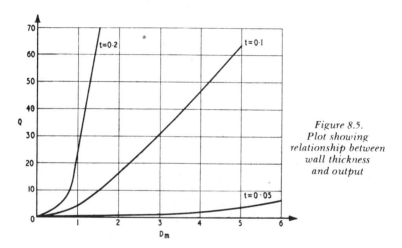

Figure 8.5. Plot showing relationship between wall thickness and output

Thus for a given pressure and ignoring considerations of non-Newtonian flow, the output from a tube die is:

Approximately linear with diameter.
Approximately proportional to the cube of the wall thickness.
Inversely proportional to the channel length.

In later work a computer has been used to assist in the design of extrusion dies and some details have been given[5]. Also, an interesting analysis, both practical and theoretical, has been carried out by Mennig into the temperature distribution of polymer melts in dies, and throws some light onto this rather obscure matter[6].

8.2.2 LAND LENGTH

In practice the wall thickness and the tube diameter are usually fixed leaving only the land or channel length to be determined by the designer; and practical experience has to some extent determined this factor also in the form of the rule of thumb land length/orifice thickness ratios which are widely used.For example a ratio of 10 : 1 is often specified for materials of high viscosity such as unplasticised PVC whereas longer lands up to a ratio of 30 : 1 may be used for low density polyethylene.

Although it is instructive to calculate the approximate flow through an extrusion die using the foregoing formulae, it must be pointed out that the usefulness of this exercise is confined to comparative values only. There are usually far too many imponderables to allow the accurate calculation of die dimensions from theoretical considerations alone and practical experience must, in most cases, be the guide at least in the early design stages. The die-land/orifice thickness ratios referred to above for example take into account, according to the experience of the designer, such considerations as surface finish, compound lubrication and filler loading, the construction of the die upstream of the land and the system of product sizing to be used, to mention only a few. It is obviously difficult to describe such factors in mathematical terms so that in the final analysis the die land/annulus thickness and other ratios established over the years by trial and error usually form the basis of most practical die designs.

The calculations are, however, often of great value when changes or modifications are required. It has been inferred above that the pressure drop across the lands of a die is proportional to the cube of the thickness: thus if the wall thickness of a tube is changed to produce a product of different dimensions, the land must be changed by the cube of this difference to maintain the same pressure gradient at the same melt temperature or viscosity.

Similarly, in considering questions of wall thickness adjustment to obtain tube concentricity the cubic factor is again important. A very small difference in die annulus width from one side of a tube die to another will result in an apparently disproportionate lack of tube concentricity. This extreme sensitivity, which is well known to all extrusion engineers, is a direct result of the cubic relationship connecting output with annulus width.

In applying the cubic relationship, it is also important to realise that with dies of very small wall thickness such as are used for the production of polyethylene film, for example, the shear rates engendered in the narrow annulus may be quite high. Besides causing a considerable local decrease in viscosity and thus of pressure drop values, these high rates of shear could also cause a considerable departure from Newtonian flow on which the cubic relationship is based. Moreover the rigorous application of the cubic relationship to the design of dies for very heavy sections on the other hand would result in very long dies which could tend to become prohibitive as the thickness increased. But provided these two extreme conditions are dealt with in an intelligent manner, the cubic expression for land length/annulus width relationship can form a very useful guide in many aspects of die design.

8.2.3 MELT FRACTURE

It sometimes occurs when operating an extruder at a high through-put rate that the extrudate takes on a rough irregular appearance which cannot be attributed to any other cause than the physical breakdown of the melt or 'melt fracture'. This phenomenon occurs when the shear stress of the melt exceeds its shear strength such as for example in an extrusion die where, due to a substantial reduction in channel width, a sudden increase occurs in the shear rate. Original investigations by Tordella[7] showed that melt fracture occurred at a critical pressure point which varied with the viscosity of the melt, the die pressure and the die design geometry. As the last factor was the only one which could be modified without causing a loss of output, the effect of the inlet geometry, land length, etc. of dies was considered. The theoretical aspects of melt fracture have also been studied by Schulken and Boy[8] and applied to the development of a method of calculating the die entrance geometry. As Tordella called it a critical pressure, so Schulken and Boy call it a critical shear rate which is dependent principally on the die entry geometry and viscosity.

8.2.3.1 *Entry geometry*

By testing each resin with a standard set of capillaries having different entrance angles and plotting the critical shear rate against the half angle of entry, *Figure 8.6*, a curve for each resin is obtained, which can be used as a reference for calculating entrance geometry from

$$\dot{\gamma} = \frac{4Q}{\pi r^3}$$

where Q is the desired rate of flow, r is the capillary radius and $\dot{\gamma}$ is the shear rate.

If we assume that two taper drills of 15° and 2½° half angle size are available and the exit diameter is known, then the flow rate Q can be calculated from the formula

$$Q = \frac{\dot{\gamma}\pi r^3}{4} \qquad (8.4)$$

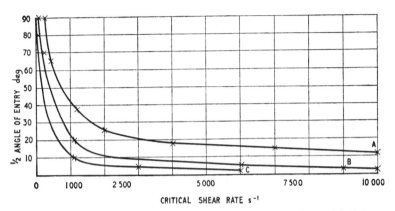

Figure 8.6. *Typical curves showing effect of capillary entry angle on critical shear rate for three different resins*

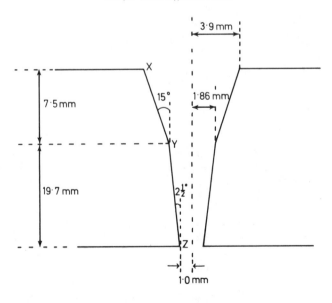

Figure 8.7. *Diagram of entry geometry*

The exit diameter of the 15° section (diameter at position Y in *Figure 8.7*) can be calculated from

$$r = \left(\frac{4Q}{\pi\dot{\gamma}}\right)^{\frac{1}{3}}$$

It is now possible to calculate the length of the 2½° section and of the 15° taper and to draw the entry geometry (*Figure 8.7*). For example, with resin B in *Figure 8.6* the exit radius (at position Z in *Figure 8.7*) is 1.0 mm the half angle is 2½°, *and* $\dot{\gamma}$ is 9000s⁻¹, giving a value for Q of

$$Q = \frac{9000 \times \pi \times (10^{-3})^3}{4} = 7.07 \times 10^{-6}\,\mathrm{m}^3/\mathrm{s}$$

Then at position Y, where the half angle is 15°, $\dot{\gamma}$ =1400s⁻¹ and Q= 7.07 x 10⁻⁶ m³/s, the radius of the entrance to the capillary is given by

$$r = \left(\frac{4 \times 7.07 \times 10^{-6}}{\pi \times 1400}\right)^{\frac{1}{3}} = 1.86\,\mathrm{mm}$$

and the length of the capillary section can be calculated,viz.
    Length = cot 2.5° x (1.86 − 1.0) = 19.7 mm
At X, with half angle 90°, $\dot{\gamma}$ = 150 s⁻¹ and Q=7.07 x 10⁻⁶ m³/s, the radius is

$$r = \left(\frac{4 \times 7.07 \times 10^{-6}}{\pi \times 150}\right)^{\frac{1}{3}}$$

and

    Length = cot 15° x (3.9 - 1.86) = 7.5 mm

The importance of entry geometry is not confined to melt fracture problems. Melt expansion caused by elastic recovery of the melt as it leaves the die is important in establishing die design. By calculating the volume of the taper and dividing it by the flow rate the melt residence time can be obtained which when compared with the relaxation rate of the melt will indicate whether expansion on leaving the die is likely to occur and whether draw-down for thickness control is called for.

The importance of die entry geometry and confirmation of its influence on extrudate quality has also been given by Metzner *et al*[9] in a study of the steady-rate flow properties of molten polymers. It is also proved that a longer-capillary die appreciably increases the stress figure at which a deterioration of quality commences.

8.3 Practical die design

Practical die design is concerned with the construction of dies for commercial extrusion in which such factors as die cost in relation to the length of run, ease of construction of the die and of adapting it for the production of other sections when it becomes redundant, ease of dismantling for cleaning and cost of replacement of component parts, are of great importance.

Moreover the ease of handling of an extrusion die and of effecting the necessary adjustments such as concentricity in a tube die, are likely to have as great or greater effect on the saleable qualities of an extrusion than is strict attention to the theories of flow.

In fact the question of die cost in relation to length of run is so important that many extrusion firms commonly produce dies for difficult profiles from flat plate with little attempt at streamlining or study of entry geometry. They are well aware that degradation of the polymer will occur at stagnant areas but such is the thermal stability of many of the present materials that the required run of profile is finished—or nearly so—before this degradation becomes important. In any case with such a simple die construction it is usually an easy matter to dismantle it for cleaning and restart[10].

In the case of dies for pipe, tubular film, sheet and other products which are substantially standard as regards dimensions, then the principal requirement is for long runs without cleaning, and this requires strict attention to streamlining and flow, which are discussed in greater detail in a following section.

In considering further this question of degradation of the polymer, a factor arises which concerns particularly the extrusion of PVC materials. This is the corrosive effect of PVC decomposition products, which can easily destroy the surface finish of an steel die, as discussed elsewhere, and has led to the use of stainless steels or to the chromium plating of the die flow surfaces for use with such materials.

An interesting alternative method of coping with this corrosion attack has recently been suggested and this is the use of anodised aluminium in extrusion die construction[11]. Such dies are easy to machine, give good resistance to PVC decomposition products, owing to the aluminium oxide skin resulting from anodising, and good service life is expected. If desired the anodised surface can be impregnated with PTFE to improve the melt flow but care is necessary in handling such dies and steel backings are often necessary.

8.3.1 DIE RESTRICTION

Figure 8.8 is a diagrammatic illustration of a typical extrusion die for the production of tube with the various important features labelled for ease of description.

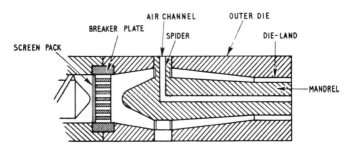

Figure 8.8. *Schematic illustration of a typical tube die*

Proceeding downstream, the die is seen to consist of a breaker plate supporting a screen pack whose fineness is determined by the material and the operating conditions, a mandrel supported by a spider structure and an outer die ring which is usually, although this is not shown, capable of lateral adjustment relative to the mandrel to obtain concentricity of the final product. The final parallel portion of the mandrel, together with the outer die ring, forms the land or annulus of the die.

In addition to being called upon to give an extrusion of predetermined form and dimensions, the die in its several parts is also required to cause resistance to flow and thus to build up back-pressures in various parts of the system.

The breaker plate/screen pack ensures pressure in the screw of the extruder, thus enabling the material to be worked and sheared and to be properly homogenised. The same effect can be obtained by the use of adjustable conical seatings at the end of the screw or by means of special restrictor valves, as is explained elsewhere. The back-pressure due to the die land restriction is relied upon to compact the material downstream of the breaker plate into one homogeneous stream.

In the case of straight-through tube dies or other extrusion systems involving obstructions to the material flow, it is also necessary besides compacting the melt into one stream, to even out the erratic velocity profiles which these obstructions cause, *Figure 8.9a*. If the restriction downstream of the spider or other obstruction is insufficient or incorrectly placed, the inequalities in velocity result in longitudinal imperfections in the finished extrusion, known as 'spider lines', which besides spoiling its appearance are potential weak lines along which the finished product could split in service. The die land restriction alone is usually insufficient to overcome such velocity disturbances and additional means are frequently provided. Many tube dies for example have a bulge built into the mandrel to form a local restriction downstream of the spider as shown in *Figure 8.10*. Correctly positioned and dimensioned, such devices can be extremely useful in overcoming these disturbances and furthermore the bulge can be shaped by trial and error to offer less restriction at positions directly in line with the spider legs or other obstructions, as shown schematically in *Figure 8.9b*.

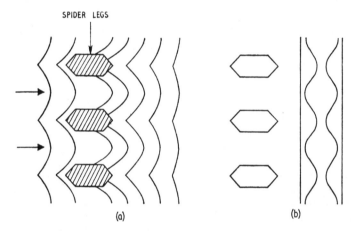

Figure 8.9. *Velocity profile at mandrel support spider*

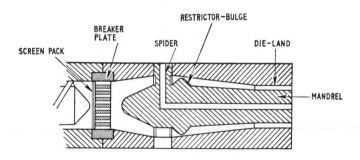

Figure 8.10. *Schematic illustration of tube die with restrictor ring*

Although in the foregoing brief reference to various forms of die restriction the matter has been treated in a general manner without reference to the material being processed, it is obvious from the expressions given previously that the melt viscosity must have a considerable bearing on the dimensions of restrictive systems. A restrictor of a certain value, i.e. the width and length of the annular gap between the bulge on a mandrel and its adjacent surface for example, which performs well with a high viscosity melt such as unplasticised PVC is likely to have very little effect if used unchanged with a low density polyethylene material at the same rate of throughput. Restrictor dimensions therefore are made adjustable or interchangeable where possible.

Finally, although this does not directly concern this matter of restriction, it is perhaps correct to refer again at this stage to the question of relaxation which has already been discussed briefly under 'melt fracture' and which is an important factor in the positioning of die parts including restrictive systems.

It is well known that polymer melts, besides in most cases being highly viscous, are also highly elastic. It is this melt elasticity which causes swelling in the extrudate as it leaves the die and is also responsible in part for the spider lines referred to above. The polymer melt having been subjected to different shear rates across the erratic velocity profiles between the obstructions arrives at the die orifice in a non-uniform state which the die land cannot overcome. Besides building up back-pressure by the use of restriction systems, it is also necessary therefore to allow sufficient time for the polymer melt to relax during its passage through the die channel after passing around an obstruction. This means that the die orifice must be placed well forward of the spider or other obstruction and it is often stated that in tube dies this dimension should be equal to at least twice the diameter of the spider. Obviously the actual minimum distance between the obstruction and the die land will be determined by the characteristics of the obstruction, the linear velocity of the melt through the channel downstream of the obstruction, and the characteristics of the polymer melt.

An ingenious way of introducing much greater length into the channel of the die downstream of the spider system without at the same time unduly increasing the physical size of the die is the subject of an interesting patent. The method described provides three interconnected concentric tubular channels through which the melt is caused to pass in succession after leaving the spider assembly, thus giving approximately three times the channel length available in a die of orthodox construction[12].

8.3.2 THE STREAMLINING OF EXTRUSION DIES

Many thermoplastics materials are heat-sensitive and too high a temperature or too long a residence time in an extruder will cause them to degrade. The control of the temperature of a polymer melt is comparatively straightforward, but the residence time is largely dependent on the design and construction of the extrusion equipment including the screw termination, the breaker plate and screen pack, the die adaptor system and the die itself.

Previously in this chapter and elsewhere[13] it has been shown that the velocity profile of the flow of a polymer melt in a channel is parabolic. Thus the material at the centre of the channel moves at a higher speed than that at the walls. In terms of an extrusion system therefore this means that there is similar distribution of residence time within the die and adaptor flow paths. The material adjacent to the walls remains in the die system for a substantially longer time than that at the centre, as can be readily observed during colour changes.

A further result of this phenomenon is that owing to the slowness of movement of the layer of melt which is in direct contact with the die or adaptor or other walls, there is a pronounced tendency for it to adhere to the metal surfaces and eventually to degrade with dire consequences to the product. In order to minimise this danger it is desirable to design the die with narrow channels—to minimise the effects of the residence time distribution gradient—and to avoid all steps and shoulders and other obstructions to flow near the channel walls. All changes of shape and dimensions must take place gradually and be smoothly blended and the internal surfaces must be highly polished and protected against the corrosive effects of the melt or its degradation products as discussed previously.

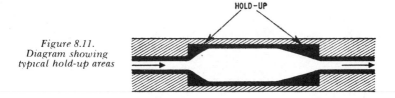

Figure 8.11.
Diagram showing
typical hold-up areas

Figure 8.11 is a diagrammatic illustration of hypothetical entry and exit regions in a channel, consisting perhaps of a die and its adaptor, with no attempt to streamline the flow. The hold-up

areas where degradation would occur are plainly seen. To minimise this effect and to achieve a reasonably trouble-free flow
through these widely differing channel dimensions it would be
necessary to design the entry and exit regions on the basis of
included angles of not more than 60° at the changes of section, as
shown in *Figure 8.12*. Ideally it would be even better to adopt the
more complicated rounded design shown reversed in *Figure 8.13*.

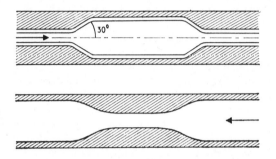

Figure 8.12.
Exit and entry
configurations
with 60° angles

Figure 8.13.
Better exit and entry
configuration showing
radiused dimension
changes

In addition to the obvious need to avoid steps and shoulders due
to badly fitting adaptors, spiders and other die components, the
breaker plate requires special attention not only as to its design
and construction but also to ensure a proper streamlined relationship with the other parts of the equipment with which it is in contact.

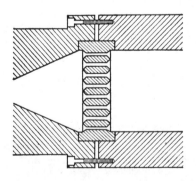

Figure 8.14.
Schematic illustration showing
a correct breaker-plate assembly

In most extruder designs the breaker plate serves as a sealing
means between the extruder barrel and the die adaptor in addition
to its other functions. Great care in design and construction is
necessary, however, to prevent the breaker plate from also acting

as an efficient means of material hold-up. The sealing surfaces of the breaker plate must be free from burrs and truly flat and in tight pressure contact with both the die adaptor and the extruder barrel. The holes must be small—3 mm (⅛ in) is usual—closely spaced and adequately streamlined or countersunk both fore and aft so that dead spots between them are avoided. Special attention is necessary to the outside ring of holes, which should be finished off with a fine file to come as near as possible to the edge.

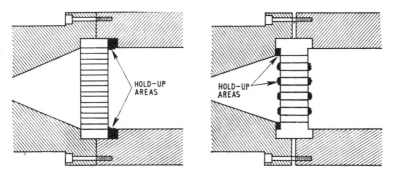

Figure 8.15. Schematic illustration showing two faulty breaker-plate assemblies

Figure 8.14 shows a well designed breaker-plate system, whilst *Figure 8.15* shows two of the faulty designs which are frequently encountered, the hold-up areas being shown by shading.

8.3.3 SCREEN PACKS

The considerable theoretical work which has been carried out on extrusion systems generally has reduced the importance of screen packs in present-day extrusion with modern equipment. In the past with deeply flighted short screws it was always the practice to build up the screen pack with successive thicknesses of fine mesh stainless steel gauges in order to reduce screw pulsation or surging and to build up back-pressure. With present-day equipment this is no longer so necessary and the main and still essential use of the screen pack is to filter out foreign matter from the material.

However, although it is now possible to design screws which in combination with a specific die design for one product or range of similar products will be quite surge-free, there are many extrusion firms whose manufacturing programme is so wide and variable that such conditions cannot apply. With such firms the screen pack remains an important means of developing back-pressure in their machines, and they commonly use a combination of 100 or even 200 mesh gauges backed up by a coarser mesh on the downstream side to prevent rupture of the finer screens. In the extrusion of dry blends a screen combination using up to twenty 325 mesh screens has been reported[14].

Better methods of developing back-pressure in the imperfectly matched screw and die combinations which frequently obtain in normal extrusion are the restrictor valves referred to earlier and described in Section 3.10 dealing with valved extrusion systems. Great care is necessary however in the design of such devices so as to avoid hold-ups, particularly in the partly closed position.

8.3.4 MEASUREMENT OF MELT PRESSURE AND TEMPERATURE

Although ample means were always provided on an extruder for heating the thermoplastics materials and converting it into a melt, it is only of recent years that the practical value of measuring the melt temperature, as against that of the barrel or die exterior, and also its pressure, has become apparent. An accurate indication and record of these values is of great assistance in obtaining optimum extrusion conditions and in ensuring reproducibility of such conditions in day-to-day operation.

For the measurement of melt temperatures in the die system it is usual to incorporate a thermocouple through the barrel wall upstream of the breaker plate or in the die adaptor immediately downstream of that member. The thermocouple should penetrate well into the melt flow, to at least a depth of 12 mm (½ in), so that its sensing tip is not influenced by the temperature of the metal barrel or adaptor through which it passes. This matter has been further investigated and discussed[15].

The most commonly used pressure measuring devices are ordinary Bourdon-tube-type gauges filled with silicone grease and equipped with grease-gun adaptors at the dial end for filling and for clearing obstructions. In order to minimise grease contamination of the melt it is usual to locate such gauges in the underside of the equipment. Pressure measuring devices may be positioned similarly to the melt temperature thermocouples, i.e. immediately upstream or downstream of the breaker plate. Both positions have their own advantages.

A better method of pressure measurement is to use a strain-gauge-type pressure transducer[16-18], which although considerably more expensive than the simple Bourdon gauge, are generally more reliable in that they seldom become obstructed with solidified material and avoid all danger of grease contamination.

8.3.5 GAUGE CONTROL

With most extrusion dies and extrusion processes the problem of control of local thickness variations in the product arises and is always difficult requiring frequent checking and considerable operator skill.

Dies for tube, for example, and for tubular film production also, have centring bolts which by displacing the outer die ring, or even in some cases the mandrel itself, can adjust the concentricity of the extrudate. Similarly sheet and flat film dies are provided with a large number of push/pull adjusting bolts which by distorting the flexible die lips or by moving a choker bar, or both, can in theory also control the local thickness of the extrudate. Such adjustments are not easy, however, and frequently cause more 'headaches' than anything else in a normal extrusion operation.

With the advent of computer-controlled extrusion processes, however—which are referred to elsewhere in this volume—it is logical to visualise thickness controls which are automatically operated by signals arising from indications which continually monitor the extrudate dimensions. This idea has been considered many times over the years in one form or another but is very difficult to implement since hitherto it has involved the concept of a 'wandering spanner' to automatically select and adjust, in or out, the appropriate bolts to apply the necessary correction.

A new system has been recently announced, however, which may in due course overcome the above difficulty and ease considerably the lot of the extrusion operator. This idea, which is known as the Welex/LFE autoflex system, uses long individually heated adjusting bolts which expand or contract to restrict or increase locally the die gap according to the temperatures of the individual bolts. By means of a closed-loop computer linked to signals arising from a continuously scanning beta gauge, a heater power source and an anticipatory memory module, the bolt lengths and thus the lateral extrudate thickness can be adjusted automatically.

The system also allows for automatic adjustment of take-off speed to control the overall thickness in the machine direction and provides a visual display of the extrudate thickness profile[43].

8.4 Typical extrusion dies

Extrusion dies may be attached to the extruder in three different ways according to the requirements of the complete extrusion process of which they form part. These three systems are known as straight-through, crosshead and offset respectively, depending on the direction of the resulting extrusion and take-off relative to the direction of melt feed from the extruder.

Straight-through dies are obviously those dies whose axes are arranged to be in line with the direction of supply of melt. It is necessary to note here that this does not necessarily mean directly in line with the axis of the extruder because in some extrusion systems a straight-through die is attached to a curved feed channel to change the direction of take-off.

Straight-through dies are commonly used for the extrusion of pipe, rod, profiles, and sheet, and, by means of a curved feed conduit, of tubular and flat film. A predominant and distinguishing characteristic of straight-through die systems is that some form of spider mandrel support assembly is essential in the production of tubular extrusions. Normal straight-through dies are shown in *Figures 8.8, 8.22,* and elsewhere.

Crosshead dies are arranged with their axes at an angle to their feed supply usually 90° but 45° and 30° are also used. Dies of this form are generally used for the production of insulated wires and cables, or in other processes where it is necessary to introduce a continuous filament, or other member to be covered, to the die mandrel. An outstanding advantage of crosshead-type die assemblies is that by this means it is possible to have ready access to the upstream end of the die mandrel so that heating or cooling or other control or manipulation of this member is easily effected. Another advantage is that owing to the use of side feed there is no need for a spider assembly in the production of hollow extrusions so that the problem of memory lines is avoided. The unbalanced feed does, however, introduce its own problems as are described later. Typical crossheads are illustrated in *Figures 8.16, 8.17a, b, d, 8.26,* etc.

Offset dies have been developed from crossheads to combine the advantages of this form of side-feed die assembly with those of the straight-through type. In this arrangement the material is made to change direction twice in an attempt to compensate as far as possible for the imbalance resulting from the single direction change which occurs in the ordinary crosshead. In this form of die extrusion usually but not essentially takes place in line with the machine. Offset dies are popular for the production of pipe where the lack of a spider and also the ease of applying temperature control to the mandrel do much to improve the quality of the product. Other advantages of the offset die system are discussed in the section dealing with product sizing. *Figure 8.27* shows a typical offset die.

Although the systems of die attachments have been classified above into three relatively clear-cut categories, each presumably being suited to one general type of extrusion, it must be pointed out that this is not strictly true and that considerable overlapping occurs. In fact as the terms straight-through, crosshead and offset refer only to the angular disposition of the die body in relation to the direction of feed, it is apparent that all three systems may be adapted for the production of any type of extruded product. The decision as to which type is to be used in a particular case is usually determined by questions of convenience, space, the availability of equipment and the previous experience of the operator. It is possible for example, but inconvenient, to use a straight-through die system for wire covering, but an offset system would be easily adaptable to this work. Similarly pipe, profiles and other products can be extruded without trouble on crosshead systems. Apart from wire-covering dies, which are now almost invariably of crosshead form, the various dies in the descriptions which follow are mostly available in forms suitable for all three die systems.

8.4.1 WIRE-COVERING CROSSHEAD

A section through a typical wire-covering crosshead is shown in *Figure 8.16* and the perpendicular disposition of the covering die with respect to the direction of the material feed can be clearly seen. Other wire-covering assemblies in which the crosshead is disposed at angles of 45° or even 30° are also used, the purpose being to minimise the flow disturbances due to the change of direction of the material whilst still retaining the advantages of the crosshead arrangement.

In the arrangement shown, which is one of many in common use, the molten material is forced through a screen-pack/breaker-plate assembly, then through an orifice which conducts the material to the closely fitting wire guide mandrel or 'point'. This mandrel is so shaped that the material flows around either side of it and joins again on the side remote from the supply. The complete annulus of material then flows towards the die orifice and ultimately contacts the wire and the forming land. In this way a tube of plastics material is deposited over the wire which moves continuously through the crosshead and acts as an internal forming mandrel.

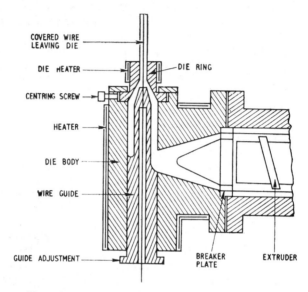

COVERED WIRE
LEAVING DIE

DIE HEATER — — DIE RING

CENTRING SCREW

HEATER

DIE BODY

WIRE GUIDE

GUIDE ADJUSTMENT — — BREAKER PLATE — EXTRUDER

Figure 8.16. Schematic diagram of wire-covering crosshead

In addition to changing the dimension of the die orifice, the thickness of the plastics coating on the wire can be affected by changing the relative speeds of the extrusion and the wire travel, as well as by changing the position of the wire guide with respect to the die ring. Provision is usually made to facilitate this last-mentioned adjustment whilst the machine is running which can also be used to affect the type of covering produced[19]. A further method which is sometimes used for sheathing bunched wires and for covering of other irregular cores is known as the 'tubing' system, where a tube is extruded from a separate orifice concentric with the wire guide and drawn into close contact with the cable by means of vacuum applied at the rear end of the guide tube. Whatever system is used, it is necessary to provide some means of ensuring that the plastics coating is concentric with the wire, since such concentricity cannot be guaranteed merely by wire and the die ring symmetry. The uneven flow conditions engendered by the method of feed of material to the wire guide and the inability to ensure even heating around the crosshead itself make the construction of a fixed die which will produce a concentric covering extremely difficult. In *Figure 8.16* it will be seen that the die ring is separate from the die body and that its position relative to the wire guide can be changed by means of centring screws. In spite of the difficulty of constructing a fixed die to give a concentric covering, many such dies are in fact used with success in the smaller sizes and are even preferred by some operators.

In order to accommodate different wire diameters and covering thicknesses, a series of wire guides and die rings are made, each having a different bore, for each of several sizes of die body. Thus each size of die body or crosshead can be used to cover a wide range of wire diameters in many coating thicknesses.

One difficulty associated with crosshead-type dies concerns the even supply of heat to the plastics material passing through the die. As can be appreciated from *Figure 8.16* heater bands can be positioned right around the die body at its two extremities, but the heat supply and radiation losses are difficult to accommodate at its junction with the extruder barrel. Sometimes an incomplete heater band is attached to one part of the exposed surface, or rod heaters may be used; sometimes lagging is used to minimise radiation losses.

Another difficulty concerns the flow of material around the wire guide. Unless this member is very carefully designed there will be a tendency for the material to stagnate and decompose at the side remote from the extruder. A number of ingenious designs have been produced to minimise this trouble[20].

A final reference to the heating problems concerns the special provision to heat the die ring which may also be seen in *Figure 8.16*. Since this part is separate from the crosshead body and moreover has a clearance around it to allow for centring adjustment, its conducted heat supply is limited to that crossing the small surface of contact between itself and the die body. In order to ensure a more efficient heat supply, therefore, a small heater is often provided solely for the die orifice portion.

Temperature control on the wire-covering crosshead is usually effected by thermocouples at two or more positions, one of which is located as near as possible to the die orifice whilst the others control the average temperature of the die body.

8.4.2 DIES FOR TUBULAR FILM

In the tubular film extrusion process a thin-walled tube is extruded and continuously inflated to a larger diameter whilst being drawn off at a relatively high linear speed. The tube is finally flattened between nip rolls to become a continuous double-thickness sheet joined at both edges. Diagrammatic illustrations of several of the many available types of tubular film dies are given in *Figure 8.17*, and a photograph of a crosshead-type die is shown in *Figure 8.18*.

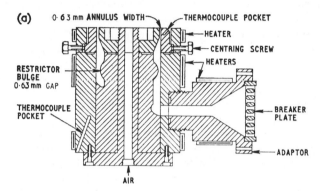

Figure 8.17a. Crosshead-type die for tubular film

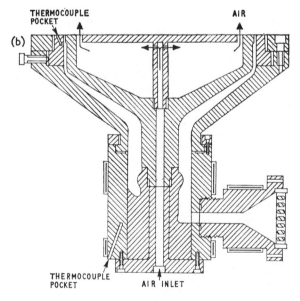

Figure 8.17b. Adaption of die shown in Figure 8.17a for large-size tubular film

Figure 8.17a is a crosshead type and is seen to be broadly the same as the wire-covering die described above. In both cases the material is supplied to one side of the die, its direction of flow is turned through 90° and at the same time it flows around a centre mandrel to form a tube whose size is determined by the die orifice. Air for inflation is supplied to the die at the end remote from the orifice through a stopcock and the heating requirements and problems are the same as those outlined for the wire-covering die.

Figure 8.17b is a tubular film die of similar type of that shown in *Figure 8.17a*, but conveniently adapted for larger sizes. *Figure 8.18* is a photograph of a typical die of this form.

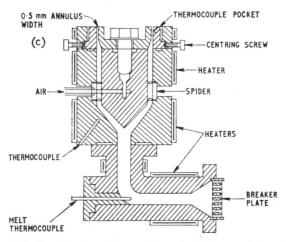

Figure 8.17c. Straight-through type (bottom feed) die for tubular film

Figure 8.17c shows a tubular film die of the straight-through type or bottom feed as it is normally known, and it will be seen that although the feed of material is symmetrical about the die orifice, thus overcoming the problem of imbalance, the new difficulty of memory or 'spider' lines due to the mandrel support system arises.

Figure 8.17d shows yet another form of tubular film die known as the 'Banjo' type, which is sometimes used for very large diameters. This die, as will be plainly seen, is again of the crosshead form and is in essence a manifold-type flat film die which has been formed into a circle. The material feed from the extruder passes directly into the circular manifold and the narrow feed channel includes a number of restrictor bulges to even out the flow around the circumference of the die lips. Such restrictions are very important with this type of tubular film die because of the very long flow path and pressure drop differences which exist between the feed side of the die as against that remote from the extruder. Large dies of the Banjo type are sometimes attached to two or more extruders by means of symmetrically arranged feed positions, to produce exceptionally wide film for use in agriculture, etc.

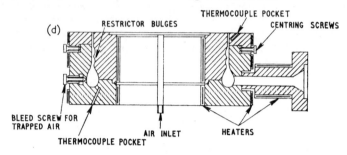

Figure 8.17d. Banjo-type die for tubular film

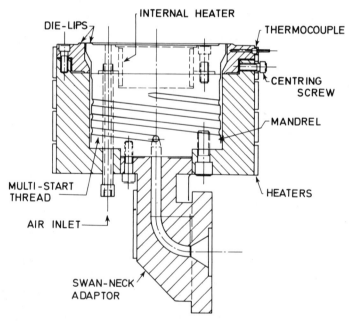

Figure 8.17e. Schematic illustration of a tubular film die with optical mandrel

Finally *Figure 8.17e* gives a schematic illustration of a tubular film die with a bottom feed but with a spiral distribution mandrel which avoids completely the two conflicting problems of imbalance due to asymmetrical feed with side-feed dies and of spider lines from the mandrel support with conventional bottom-feed dies. Such die systems, although ideal in many respects, do frequently introduce their own problems of excessively high die pressures and long transition times and heat history. It will be noticed in the illustration, which is somewhat exaggerated, that the molten material is forwarded to the die land both along the spiral channel and over the flights thereof. By the careful balance of these two flow systems uniform feed to the land can be obtained and the pressure and residence time adjusted. See also *Figure 8.19*, which is a photograph of such a spiral mandrel die in dismantled form.

Figure 8.18 A typical crosshead die for tubular film. (Courtesy A. Reifenhauser, Troisdorf)

Most tubular film dies of whatever type are provided with means for adjusting the concentricity of the tube, either by moving the inner mandrel or, more often, the outer die ring: or even by distorting the outer ring by a series of adjusting screws positioned at small intervals around the circumference. The importance of obtaining good concentricity in the tubular film process cannot be overstressed and there have been many ingenious attempts to simplify this rather critical adjustment[21-23].

Figure 8.19 A spiral mandrel blown film die dismantled to show its components. (Courtesy N.R.M. Corpn. Akron, Ohio)

The die annulus width of normal tubular film dies is set at about 0.6 mm (0.025 in), although other thicknesses are used to obtain special properties in the film such as balanced orientation.

8.4.3 FLAT FILM DIES

In the trade it is usual to classify web products with a thickness of below 0.25 mm (0.010 in) as film whereas material above that thickness is classified as sheet. This distinction, which is purely arbitrary, is convenient because the production methods tend to change as the product approaches 0.25 mm (0.010 in) in thickness and because the product itself takes on different properties as the thickness increases.

Film products generally are extruded as melts, and then drawn down whilst still in the molten stage to 1/20 or thereabouts of the die orifice thickness; this technique is essential because of the obvious difficulty of extruding from an orifice of only 25 μm (0.001 in) in thickness for example. Sheet on the other hand is not subjected to this high draw during manufacture, which would introduce high internal stresses and give great difficulty in subsequent heat-forming operations.

Figure 8.20 is a diagrammatic sketch of a typical die, known as a manifold die, for the extrusion of unsupported flat film or for the coating of paper and other substrates. For this latter purpose it is necessary to construct the die with protruding lips to allow close positioning of the laminating rolls. Manifold dies can be used successfully with polyethylene, nylon and some other materials. The die consists essentially of a tube which is slit longitudinally, closed at the ends and provided with at least one adjustable jaw for control of flow rate and film thickness. The feed to the centre channel or manifold can be applied at one end or in the middle of the tube and provided the viscosity of the melt is maintained at a fairly low value, no undue difficulties arise in connection with uneven flow from the die orifice. All internal flow surfaces must be highly polished.

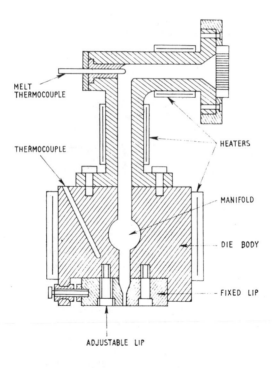

Figure 8.20. Standard manifold-type flat film die

Heaters are positioned on both flat faces of the body and it is essential that such heaters be symmetrically arranged in order to avoid warpage of the die due to irregular expansion. Heater loading of approximately 8 kW/m (200 W/in) of die length is usually required and accurate temperature control at a number of points along the whole die length is essential.

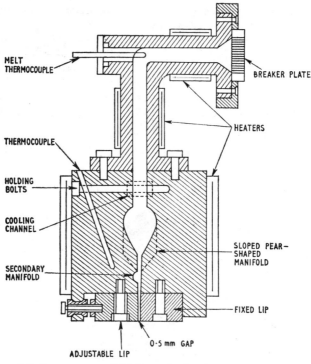

Figure 8.21. Later-type manifold die for flat film showing sloped pear-shaped manifold and cooling channel

A sectional view of a more modern type of manifold die is shown in *Figure 8.21* and a number of important improvements will be noticed. The manifold is now pear-shaped and also slopes towards the die lips as it approaches the ends of the die body. These changes give better streamlining and assist the flow at the die extremities. A second small manifold is provided to even out flow disturbances and a cooling cavity is machined throughout the length of the die body through which water can be circulated at shut down to avoid degradation and contamination of the flow path and die lips with burnt particles.

8.4.4 SHEET DIES

A diagrammatic arrangement of a typical die for the extrusion of sheet in materials such as polyethylene, polystyrene, acrylics, cellulosics, etc. is shown in *Figure 8.22*. This die, which is used for products from 0.6 to 6mm (0.025 to 0.25 in) or thereabouts in thickness, consists of a split die body with a longitudinal manifold, an adjustable restrictor bar and adjustable and interchangeable lip members. By careful adjustment of the die lips and the restrictor bar, using the setting bolts provided, the required pressure profile can be obtained, thus ensuring uniform delivery of material from the lips. The ends of the die are sealed with plates to which are attached profiled plugs which fit into the ends of the manifold and help to direct the flow towards the die lips. In some cases adjustable bleeder holes are provided in the end plates to speed-up colour or material changes.

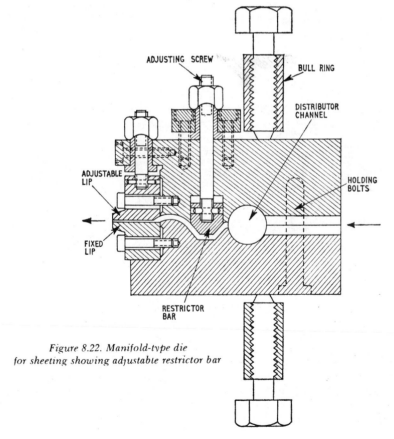

Figure 8.22. Manifold-type die for sheeting showing adjustable restrictor bar

A suitable number of resistance heaters are clamped along the upper and lower surfaces of the die and controlled in zones so that if necessary a temperature gradient can be maintained along the length of the die body to give further control of melt flow. Owing to the high separating forces which build up in dies of this type, it is often necessary to provide an external clamping system known as a 'bull ring', which consists of a heavy steel member surrounding the die from which protrude a number of large adjustable bolts to bear on the die halves and thus to take the bursting loads. Owing to the considerable weight of sheeting dies and their clamping rings, when used, it is usual to mount the whole assembly on a trolley as shown in *Figure 8.23.*

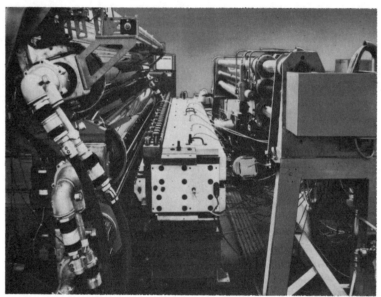

Figure 8.23 A 3.5 m sheeting die with associated equipment. (Courtesy Reifenhauser K.G., Troisdorf)

The extrusion of sheet in materials of high viscosity and particularly if they are also heat-sensitive, as for example with rigid PVC, presents considerable further problems because of the tendency of such materials to hold-up and degrade at the die extremities. The normal manifold type of sheeting die as described above is therefore not suitable for such work and other designs and methods have been developed for this application.

One form of die which is more successful in the extrusion of rigid PVC materials is known as the fishtail die. In such a die the cavity is so shaped that the distance and resistance to flow from the entry to the die lips is approximately equal across the entire die width. However, although such a design of die reduces the flow problems, it is subject to very high separating forces which are difficult to accommodate merely by bolting around the edges and which tend to force the lips apart at the central area. The use of one or more external clamps is therefore essential.

An interesting method of overcoming this problem without the use of external clamps is provided by a later design of die in which the fishtail cavity is folded over on itself so that the lips are in effect at a different level from the cavity. This die is split vertically along its length and although costly is able to contain the high bursting pressures without distortion.

Another ingenious method of achieving uniform distribution of material is to provide a stirrer or other rotating member in the manifold of an otherwise normal sheeting die[24],[25].

Although wide flat manifold- or fishtail-type dies are often looked upon as being the only methods of extruding sheet, it must not be forgotten that the earliest systems used dies of tubular form: the tube so produced was slit and flattened as a continuous operation. This method appears to have been developed in Italy during the early post-war years for the production of rigid PVC sheet and in the U.S.A. for cellulosic materials at approximately the same time[26]. In later years, however, it was abandoned for cellulosics in favour of the wide-slit die system, but is still retained in Italy for PVC materials. The tubular system in a modified form is now used also for the extrusion of sheet in expanded polystyrene.

The obvious advantage of the tubular method of producing rigid PVC sheet is the complete avoidance of the problems of flow which require much ingenuity, skill in design, and complication even to minimise in the wide-slit die systems. On the other hand, the problem of spider lines, or of imbalance if an offset system is used, arises as also does the difficulty of post flattening without introducing internal stresses.

8.4.5 TUBE DIES

Typical straight-through extrusion dies for the production of tube in rigid or semi-rigid materials are shown in *Figures 8.24* and *8.25* and dies of crosshead and offset form are shown in *Figures 8.26* and *8.27* respectively.

The straight-through die is seen to consist essentially of a die body which supports the main structure and an outer die ring and die mandrel which together give the orifice dimensions. The mandrel is supported on a multi-spoked piece called a 'spider' which fits into a recess in the die body where it can be rigidly clamped.

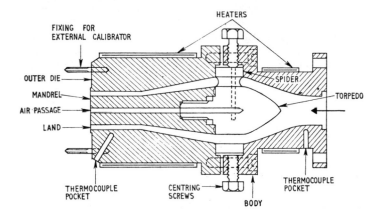

Figure 8.24. Schematic drawing of a die design for small and medium-sized tubes

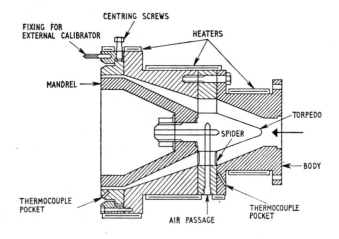

Figure 8.25. Schematic drawing of a large tube die

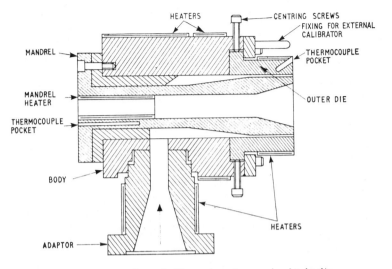

Figure 8.26. Schematic illustration of a crosshead tube die

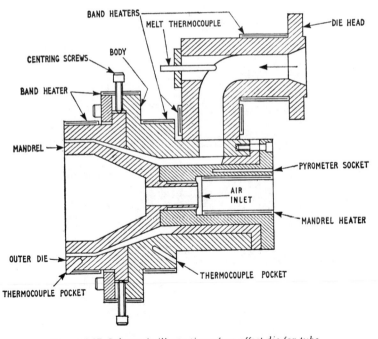

Figure 8.27. Schematic illustration of an offset die for tube

Adjustment for tube concentricity may be obtained by means of centring screws which move the spider and the attached mandrel as shown in the illustration of the smaller die or by moving the outer ring in relation to a fixed mandrel as shown in the other illustrations. In the latter system, which is the most frequently used, the adjusting screws bear on the outer die ring which is recessed into the die body and clamped in position by means of a separate clamping ring or other device. The fixed mandrel and spider give a more rigid construction which is essential for tubes of larger diameter.

An air hole is drilled up the centre of the mandrel which connects with a hole through the centre of one or more of the spider legs or spokes. This hole in turn connects with an accurately controllable and uniform supply of compressed air which is continuously delivered to the interior of the tube being extruded. The upstream end of the spider has a torpedo-shaped protrusion which guides the melt to the through passages and facilitates smooth flow. The spider is usually constructed to receive a range of mandrels of different size and thus to produce products of various bores.

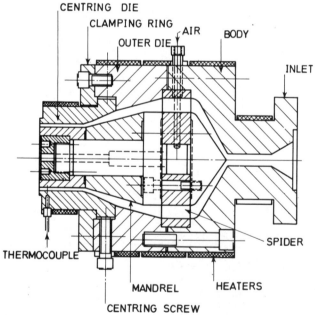

Figure 8.28. Schematic illustration of a later type of pipe die showing the circumferential compresssion effect.

Resistance heater bands are clamped around the outside of the die body and outer die and one or more thermocouples are provided for temperature control. On some large-diameter dies a cartridge heater is inserted into the mandrel in order to heat the inner surfaces of the die. A schematic illustration of another type of straight-through die now much used for the extrusion of pipes, particularly in rigid PVC and similar materials, is given in *Figure 8.28*. It will be seen that when the material leaves the screw or screws it is compressed or compacted into a rod of small diameter and is then redistributed by the torpedo member of the mandrel-supporting spider which is of diameter and cubic capacity considerably greater than that of the final die annulus. In its passage through the large-capacity spider and associated flow channels, the material is progressively compressed circumferentially into the much smaller die annulus and is thus compacted into a homogeneous product free from spider lines and other defects.

Figure 8.29 A die for medium-size pipe.
(Courtesy Cincinnati Milacron)

Pipe extrusion dies of this type are expensive to produce because. of the large spider and body dimensions relative to the size of tube produced, but the circumferential compression imparted to the material appears to be a very important factor in the production of high quality pipes for pressure application. A photograph of a die of this type is shown in *Figure 8.29* and the massive dimensions involved will be apparent.

8.4.6 DIES FOR SOLID SECTIONS

Dies for the extrusion of solid sections are similar in general conception to the dies for tube but are simpler in that there are none of the many problems associated with mandrels and spiders. Profile dies often consist of nothing more than a feed portion or adaptor which is attached to the extruder and an interchangeable die plate in which the profile is cut and blended at the entry to mate with the exit of the feed portion.

Dies for the extrusion of solid rod may be considered as tube dies from which mandrels and spiders have been omitted, and which have been given lands of sufficient length to build up the requisite back pressure.

Because of the surface drag on the material passing through the die and also because of the parabolic nature of the velocity profile of flow referred to earlier in this chapter, there is a marked tendency for greater flow to take place through the thicker parts of a profile die than through the thinner parts. The die orifice therefore has to be specially shaped to counteract this tendency and its form is seldom similar to that of the extrusion obtained from it. In *Figure 8.30* the full lines indicate the shape of the required profile and the dotted lines show approximately the form of the die orifice required to produce such a section. It will be noted that provision has to be made to augment the supply of material to the corners where the effects of surface friction from the two perpendicular faces become apparent. It is also possible to vary the land length at different parts of the die and both land length variations and orifice adjustments are frequently used in combination to obtain a required section.

It is obvious from *Figure 8.30* that a complex profile consisting of a variety of thicknesses, corners and angles could present considerable difficulty and would be impossible to design from a theory of flow. The usual method is to produce a profile plate in unfinished form and to arrive at a correct profile by trial and error. One or two interesting systems of speeding up this procedure have been suggested, however[13]. Another method which is sometimes used for intricate profiles is to cast a die form in a suitable soft alloy using a plaster positive of the required section. This cast is then retained in a suitable holder, and is used as the basis for extrusion trials. The soft metal can be easily removed if adjustments to the die form are required, and the finished soft die can be used for a short production run. If a long run is required, the soft die can be duplicated in steel.

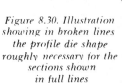 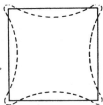 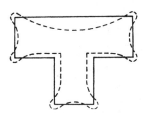

Figure 8.30. Illustration showing in broken lines the profile die shape roughly necessary for the sections shown in full lines

A certain degree of draw-down is normally applied to uncalibrated sections when they leave the die in order to maintain sufficient tension to prevent sag and as a method of obtaining the required section dimensions by adjusting the speed of take-off. The die profile must, therefore, be somewhat larger than the required section. However, this draw-down produces unequal contractions in a varied section since thick parts tend to reduce less than thin parts. A flat strip of material, for example, tends to thin out far more in its thickness than in its width. Allowances must be made, therefore, for these further section changes when designing a profile die.

Further information on the extrusion of complicated profiles is given in a later section of this chapter which deals with calibration.

8.4.7 DIES FOR MULTICOLOUR OR MULTI-MATERIAL EXTRUSION

It is obvious that if a die be constructed with two or more inlets connected to separate channels arranged to converge just upstream of the die land, and if the inlets are attached to separate extruders fed with different materials, then a multi-extrusion will result. There is nothing particularly difficult about such a die arrangement, apart from its rather complicated internal flow path structure, and if the materials used are mutually compatible then a good bond between the component parts will be achieved. It is necessary, of course, to exercise care in control of the output rates of the various feed sources so that an overall uniform rate of product output is obtained. It may also be necessary to adjust the position in the flow channels relative to the die land entry where the materials converge to obtain a compromise between good bonding and a clearly defined weld line.

Apparatus of the above type has been used for the production of multicoloured decorative strips and beadings, for the production of a flexible edge to an otherwise rigid profile and for the manufacture of multicoloured tube. It has also been used in conjunction with twisting equipment, for the manufacture of spirally multicoloured insulated wires[27] and in the production of multi-layer tubular film, flat-film and sheet in the process now sometimes known as 'coextrusion'.

Schematic illustrations of typical flat film and tubular coextrusion dies are given in *Figures 8.31* and *8.32* and further information is given in the following chapter dealing with complete extrusion processes.

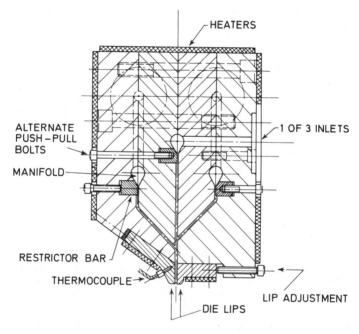

Figure 8.31. Schematic illustration of one type of flat film coextrusion die. (Exaggerated in some respects for clarity)

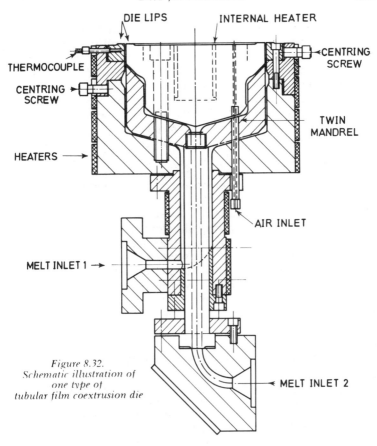

Figure 8.32.
Schematic illustration of
one type of
tubular film coextrusion die

8.4.8 MECHANICALLY DRIVEN DIES

Apart from the use of driven screws or stirrers in some part of the die in order to achieve a more uniform flow distribution, as referred to briefly in Section 8.4.4, there have been many ingenious and often successful attempts to produce novel extruded products by means of crosshead dies with mechanically driven members of one form or another. Dies with reciprocating mandrels for example have been used for the production of tube with alternating thick and thin walls, and another ingenious die used a rotating mandrel connected to two extruders fed with different coloured materials to produce a spirally variegated multicoloured tube.

Yet another device supports the die mandrel on the bore of a rotating member in which a thread of special form is cut. By this means, it is possible to feed a spiral spring through the thread of the rotating member along the mandrel in such a way that it becomes embedded in the walls of the plastics tube being extruded, to produce a steel-reinforced hose. Another but perhaps less satisfactory method of achieving the same end uses three driven spur gears arranged radially at 120° to the mandrel which they support. Three depressions are machined on the mandrel diameter at the appropriate positions where it contacts the gears and the wire spiral finds its way into the crosshead flow path via the teeth of the spur gears.

Perhaps the most elegant and successful mechanical die system of recent years is the one invented by Mercer for the extrusion of plastics net[28]. This ingenious device is in effect a multi-orifice spaghetti die containing a ring of holes which are drilled half in the mandrel and half in the outer die ring along the centre line where these two members are in contact with each other. In operation the mandrel and the die ring are rotated in opposite directions, and where the half-holes coincide, the extruded semi-circular strands, which are continuously produced, will cross over and be welded into knots.

8.5 Extrusion sizing or calibration systems

In the majority of cases an extrudate when it emerges from the extrusion die is still in a semi-molten state which, to produce a useful product, requires to be continuously chilled whilst being held to the required form and dimensions. This statement does not apply to soft or rubber-like products from plasticised PVC and similar materials; with which positive sizing systems are seldom used or possible.

In some processes such as the production of tubular film or of flat film, impingement by a large volume of cold air, plain entry into a water bath at the correct speed, or contact with a chilled roll depending on the product, is sufficient to achieve this aim. But in other processes such as rigid tube or profile production more positive methods are needed. The devices for this purpose are known variously as calibrators, forming boxes, sizing formers, sizing dies, etc., and in some cases might be considered as part of the take-off system. But as they are usually attached to, or closely associated with, the extrusion die, it seems correct to discuss them in this chapter.

8.5.1 CALIBRATORS FOR RIGID AND SEMI-RIGID TUBE

There are two major systems of calibration in common use for tube, i.e. the external method employing a water-cooled forming tube in conjunction with internal air pressure, and the internal system employing an extended cooled mandrel and secondary external cooling. The external system is mostly used in Europe and the U.K., whilst the extended mandrel method, which is suitable only for crosshead or offset dies, was popular in the U.S.A. but is becoming less so with the importation of European twin-screw technologies.

Tube production without any form of calibration, however, is still practised on a limited scale for the manufacture of products where a degree of ovality is permitted; such as for example, in the production of tube which is to be post-formed or subsequently split and opened into sheet. Rigid PVC pipes up to 750 mm (30 in) in diameter and larger have been produced in Italy by such means.

8.5.1.1 External calibrators

The general form of most normal external calibration devices is shown in *Figure 8.33*. Such calibrators, which were first used by the author in 1945, and thought to be original, consist basically of a water-jacketed tube, the bore of which is machined to the required outside diameter of the tube to be produced plus a calculated allowance for shrinkage during final cooling. This assembly is mounted in exact alignment on the front of the extrusion die which may be of the straight-through, crosshead or offset form, in such a way that heat transfer between the two members is minimal.

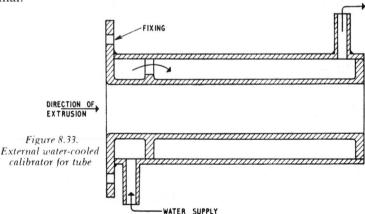

FIXING

DIRECTION OF EXTRUSION

Figure 8.33. External water-cooled calibrator for tube

WATER SUPPLY

In operation the soft thermoplastics tube with its end sealed is passed through the calibrator and continuously inflated via the drilled die mandrel so that its outside surface is in sliding pressure contact with the chilled interior surface of the calibrator bore. The air pressures used depend on the diameter and wall thickness of the tube and the viscosity of the material, and vary between 3.5 and 275 kN/m² (½ and 40 lbf/in²). In this manner the outside layer of the tube is frozen to size to become self-supporting whilst final cooling throughout the wall thickness is then given by subsequent passage through a normal water bath or by water spray. *Figure 8.34* is an actual photograph of a 300 mm (12in) bore, 9.5 mm (⅜ in) wall tube in melt index 20 polyethylene being produced in 1947 on equipment of this type designed by the author.

Figure 8.34 A 300mm (12in) diameter pipe being produced in 1947 using external calibration. (Courtesy Tenaplas Ltd)

The length of the calibration tube must be such that the thermoplastics tube is frozen on its outside surface during the contact period to a depth sufficient to prevent distortion or busting during its passage into and through the initial stages of the final cooling system. This length therefore depends on the linear throughput speed of the tube, the effectiveness of subsequent cooling and on the plastic temperature range, heat capacity and heat transfer characteristics of the material. It is always best to use the shortest possible calibrator combined with an effective subsequent cooling system rather than the reverse because long calibrators, besides being inconvenient to operate, cause high friction, introduce strain, and do not contribute a proportionate degree of extra cooling.

The internal surface finish of calibrating formers is also important. The finish must be such that it gives maximum heat transfer from the hot thermoplastic, and yet does not adhere to it or cause excessive friction. Roughened surfaces as by sandblasting or by fine machining with a single point tool give good results with many materials such as low or high density polyethylene but with other thermoplastics, i.e. rigid PVC a polished surface seems to be more effective. In the early days of the use of external calibration systems methods of lubrication were experimented with using glycerol and other fluids[29] and an air system was also developed[30]. At the present time, however, most of these innovations have been abandoned and high speed operation is achieved by the correct design of a normal cooled calibrator with precisely controlled internal pressure to work in conjunction with an efficient secondary cooling system.

All external calibration systems which rely on the use of internal air pressure to bring about sliding contact with a chilled forming tube require to be sealed at the far end to prevent loss of air pressure. In the case of rigid tubes which cannot be coiled and pinch-sealed, but have to be cut off in set comparatively short lengths, this presents a problem, for every time the tube is cut there is a danger of pressure loss and consequent fall-away of the tube from the calibration surface. It was common practice for the cut-off saw operator to rapidly replug by hand the tube immediately after the cut and considerable skill in this operation is achieved with practice. On the other hand, such a method is somewhat crude and obviously unsuited to modern automatic systems with low labour content and to large-diameter pipes. Other systems of preventing air loss have been devised, therefore, and perhaps the most commonly used of these is the floating plug attached to the die mandrel by means of a chain or other flexible member *(Figure 8.35)*. A further system uses a floating plug held in position by electromagnetic means[31] and yet another method uses screw-operated clamps which are part of the take-off and which remain with the tube as it passes away from the sizing device: a second clamp is attached at the sizing device before the tube length is cut off. This equipment is discussed further in a later section.

The most elegant method of avoiding the effects of internal pressure loss with external calibration systems relies on the use of vacuum to bring about the required sliding contact rather than positive internal pressure. This method, which unfortunately is limited as to the pressure differential available, is nevertheless extremely effective with small and medium-sized tubes in many materials and avoids all tube cut-off pressure loss problems[32].

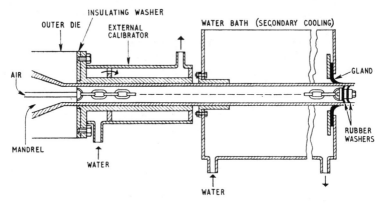

Figure 8.35. External calibration with floating plug on chain

Two basic methods of vacuum sizing have in fact been used, one is designed as a single unit where alternative vacuum/water-cooling sections control the cooling and sizing of the pipe. A second design incorporates several vacuum sizing units which are adjustable in their longitudinal position within a cooling tank. This second system is claimed to give full flexible control and to be easily adjustable. A third and simpler method which has also proved successful uses a vacuum tank and in this system the extrudate passes through some form of sizing die into a water bath which is held under vacuum.

8.5.1.2   *Internal calibrators*

A diagrammatic illustration of one form of internal calibrator is given in *Figure 8.36*, where it can be seen to consist of a tapered water-cooled extension to the mandrel which fixes the inside diameter of the tube in much the same way as external calibration systems control the outside dimension. Apart from the fact that internal calibrators are normally used with offset dies only—the supply of coolant to the mandrel extension being difficult with the straight-through type—the main advantage of this system is that it enables the tube to be cooled simultaneously on both the inside and the outside, and therefore allows higher take-off speeds than are possible with external calibration, which cools the exterior only.

Other advantages of the internal system are that all problems of cut-off such as are encountered in the external system with positive air pressure are avoided and that by this means it is often possible to produce accurate hollow profiles of non-circular form from a circular die by merely replacing the tapered cylindrical mandrel extension by one of the required form.

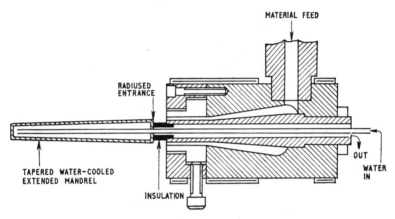

Figure 8.36. Internal calibrator with offset die

In operation the extrudate is drawn over the extended mandrel on to which it tends to shrink during cooling; this shrinkage provides the calibration pressure necessary to size the product. The take-off force required is usually higher than that necessary with an external calibration system, but the degree of shrinkage and, therefore, of the pull required may be controlled by mandrel length and taper, by the temperature of the cooling medium, and by the mandrel surface, which may be either polished or deliberately roughened. These adjustments will naturally depend on such factors as the linear output rate of tube, the initial extrudate temperature, the wall thickness and the physical characteristics of the polymer, but it is usual to start with a taper of about 0.004 millimetres per millimetre (0.004in per inch) of extended mandrel length, which would be about 230mm (9in) long for a 75mm (3in) i.d.x6.3mm (¼in) wall pipe. The mandrel temperature would be controlled at about 55°C for a medium density polyethylene and somewhat lower for a rigid PVC extrusion.

8.5.1.3 Combined external and internal calibration

Both the external and the internal methods of tube calibration have their own limitations peculiar to the system of operation. In the case of the external process the outside of the tube, being in sliding contact with a uniform metallic cooled surface, acquires a high polish, but the interior can remain rough or undulating owing to defects in the material or to the use of incorrect extrusion conditions. The interior system on the other hand gives a smooth bore to the tube, but in effect transfers any defects to the exterior.

From the point of view of service, therefore, the internal calibration system usually gives the better product as there can be no defects in the bore to impede the flow of fluid within the tube. The external method on the other hand usually gives the best looking product and is more suitable for standard fittings.

An ingenious system for combining both the external and the internal methods of calibration has been patented by L.M.P., Turin (Roberto Colombo)[33]. In this system a short water-cooled external calibrator is combined with an extended mandrel which is heated at its upstream end. Intense secondary shock cooling is therefore provided to the tube where it issues from the external calibrator and is still in contact with the heated extended mandrel.

A high finish is given to the tube exterior by contact with the external calibrator, and by means of carefully designed internal flow channels to the die the inventors claim to provide an equal finish on the interior without troubles due to sticking on the hot mandrel.

8.5.1.4 The sizing of small tube

The two systems of tube calibration which have been outlined above are suitable for tubes of medium and large size, but are not generally applicable to diameters of less than about 10 mm (⅜ inch) bore. The reasons for this are obvious. It is extremely difficult to construct an internal calibrator which would work satisfactorily on such small sizes and the external systems are also attended by practical difficulties of operation which increase as the size is reduced.

These difficulties are modified somewhat by the melt characteristics of the polymer being handled. Rigid PVC for example, can be calibrated by external systems down to quite small diameters, but the polyolefins on the other hand and other materials which are very soft in their plastic state cannot be so handled, and other systems of sizing have been devised.

The most common method for the sizing of small tubes uses an extrusion dié which is larger than the required final diameter and draws the product through a series of brass sizing plates which are contained in a water bath, the first sizing plate in effect forming part of the front end of the water bath. The distance between the die and the first sizing plate, known as the 'air-gap', is adjustable, as are the gaps between following sizing plates in the water bath. The method depends for its success on the rapid draw-off, which results from the use of an oversized die, thus relying on tension to prevent sag and distortion. The degree of draw-down required in a particular case, therefore, is determined by the melt characteristics of the extrudate. Rigid PVC for example will require very little draw, whereas some polyolefins need a draw ratio of as much as 5 : 1.

The sizes of the orifices in the draw plates are dependent on the draw ratio and also on the length of the air-gap. In an average case the first draw plate would be 10 per cent oversize, whilst the second and third plates would be size plus an allowance for shrinkage on cooling. It is usual also to bevel the plate orifices at 45° from the inside surface in order to minimise drag.

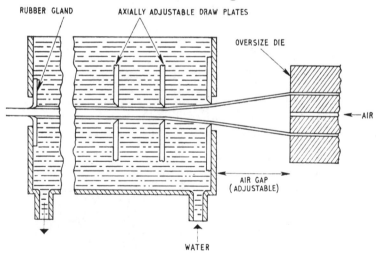

Figure 8.37. Illustration showing the sizing of small tube

In operating the above system the wall thickness of the product is controlled by take-off speed and the outside diameter by the sizing plates and it is usual to give sufficient internal air pressure to the tube so that it just fills the first die plate orifice without any pronounced bulge in the air-gap (*Figure 8.37*).

8.5.2 THE CALIBRATION OF PROFILES

Very little specific information can be given on the calibration of
profiles because as with the design of profile dies each shape presents its own problems which require individual study. Such factors as the melt characteristics of the extrudate, the thickness and
shape of the required product, and the degree of draw-down all
influence the method of calibration to be adopted. In general, however, the methods and equipment used are frequently very similar
to those adopted for small tubes. The profile is extruded from an
oversize die and is drawn through a series of sizing plates to obtain
final shape and dimensions. However, due to the extreme delicacy
of the extruded shape as it emerges from the die, it is usual to apply
some system of surface cooling in the air-gap before the product
contacts the first draw plate. This extra cooling often takes the
form of a tube surrounding the profile through which a current of
cold air is carefully passed in the direction away from the die, or
instead a fine water spray may be used.

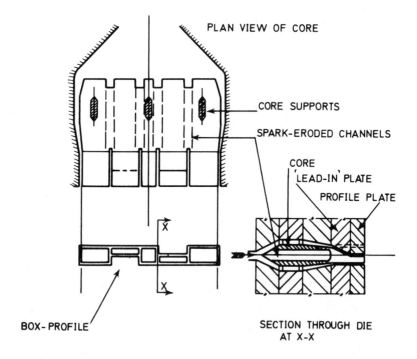

Figure 8.38. Schematic view of a profile die showing method of construction

The extrusion of complicated profiles is a highly skilled operation calling for considerable experience both in the design of the die and sizing equipment and in the operation of the machine. An extrusion with long upstanding ears, for example, might tend to collapse inward immediately on emerging from the die. In this case a stream of low pressure air might be directed on to the ears to prevent this collapse or the die might be designed with the ears diverging so that they would collapse into the required position. Other methods of controlling the form of extruded profiles use water-cooled fingers in the air-gap to hold the various projections etc. in position and yet others use water-cooled brass calibrators which are split and hinged to open horizontally and which are internally machined, with a taper to allow for shrinkage, to the exact shape and dimensions of the required profile.

By way of an example, *Figure 8.38* is an illustration of a complicated but very common form of profile of which the accurate extrusion would present great difficulty—if indeed it were even possible—without some sophisticated calibration means. It will be noticed that this particular profile consists of a number of interconnected tubes of substantially rectangular form, each of which could be individually extruded and calibrated without undue difficulty using suitable external calibrators and internal air pressure.

The complete profile would therefore be produced in a similar manner using a die designed to provide air pressure to each hollow part and a calibrator shaped to the final required profile dimension.

Further information on methods of sizing profiles and small tube is available in patent and other literature[35–36].

8.5.3 THE SIZING OF ROD

In the section dealing with extrusion dies for rod it was implied that such dies were very simple, consisting in effect of nothing more than a tube die with the mandrel and spider removed and given sufficient land length to ensure adequate back-pressure.

This is correct, but far from being simple, the actual extrusion of high quality rod, especially of large diameter, is one of the most difficult of extrusion operations.

The first problem is associated with the distribution of pressures in thick-section extrusion dies and with the frictional effects of the die walls. These factors cause the material to extrude at higher speed in the centre of the die orifice than at the outside, where it is in contact with the walls, which frequently results in an unstable, internally stressed product. In very thick sections it can also result in the material in the centre actually extruding through that at the outside whilst the latter material remains stationary or stuck to the die walls.

These difficulties may be overcome or minimised by rapidly freezing the advancing face of the rod extrudate to convert the laminar flow into plug flow, by using highly polished die surfaces to reduce friction, and by increasing the die temperature so that the viscosity of the material in contact therewith is reduced, to act as a lubricant for the main body of the extrusion. Other methods using extra long pressure lubricated dies have also been used with success[37],[38].

The second main problem encountered in the extrusion of rod is the production of voids and bubbles due to differential contraction. All thermoplastics materials have high coefficients of thermal expansion and are poor conductors of heat. Moreover, some polymers—the crystalline types—solidify more or less rapidly, in some cases within a narrow temperature range, and show substantial volumetric contraction on solidification.

When an extrudate of thermoplastics leaves its die it is substantially uniform in temperature and is homogeneous. Owing to the above properties, however, immediately it is subjected to a cooling system the exterior commences to solidify and contract into a rigid shell whilst the interior remains molten and at maximum volume. The subsequent solidification contraction of the interior within the rigid exterior shell gives rise to voids which in large sections can be of substantial size.

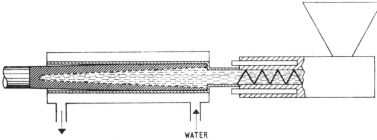

WATER

Figure 8.39. Method for the extrusion of rod in crystalline polymers of low melt viscosity

The methods adopted to overcome these difficulties depend to a considerable extent on the characteristics of the polymer being handled. In the case of thermoplastics with a wide plastic range the exterior cooling can be delayed so that the outer shell remains soft and is thus able to contract to accommodate the subsequent loss of volume on the interior. This method, which is widely adopted in the cable industry for the extrusion of thick insulation, cools the product by passage through a succession of water baths with gradually reducing temperatures, the first bath being as hot as possible, whilst the last one is at ambient. The effectiveness of this method may be improved by the application of external fluid pressure during cooling, but the system is obviously limited as to the size of product which can be handled and to the range of polymers with which it would be effective.

In the case of polymers which show substantial contraction on solidification, and which have relatively sharp melting points, and melts of comparatively low viscosity, the above methods of avoiding porosity are not effective. Such polymers give the greatest trouble in rod extrusion but fortunately an ingenious system of overcoming the difficulties has been devised, and patented. This system uses a cooled forming tube as a die and relies on the build-up of pressure in the molten core of material to expand the solidified shell as it is formed which thus acts as a braking force against the tube bore. Thus the extruded rod progresses slowly or jerkily through the tube and by means of a nice balance between rate of cooling, melt temperature and extrusion rate it is possible to preserve the pressure in the molten interior and in consequence to extrude a void-free product (*Figure 8.39*)[40]. A more sophisticated version of the same method uses constant speed take-off rolls and a system of melt pressure control in the extruder and die[41,42].

REFERENCES

1 WEEKS, D.J., 'Some Aids to the Design of Dies for Plastic Extrusion', *Br. Plast.*, **31**, 156 (1958); **31**, 201 (1958)

2 TOBOLSKY, A.V., POWELL, R.E. and EYRING, H., *The Chemistry of Large Molecules (Frontiers of Science*, Vol 1). Interscience, New York (1943), p. 125

3 BIRD, R.B., 'Viscous Heat Effects in Extrusion of Molten Plastics', **11**, 35 (1955)

4 SCHENKEL, G., 'Developments of Extrusion in Germany', *Plast. Prog.*, 187 (1957)

5 FENNER, R.T., 'Designing Extruder Screws and Dies with the Aid of Computers', *Plast. Poly*, **42**,114 (1974)

6 MENNIG, G., 'Temperature Distribution in Highly Viscous Non-Newtonian Polymer Melts Flowing through Dies', *Plast. Poly.*, **40**, 330 (1972)

7 TORDELLA, J.P., 'Melt Structure — Extrudate Roughness in Plastics Extrusion', *S.P.E. Jl*, **12**,36 (1956)

8 SCHULKEN, R.M. and BOY, R.E., 'Cause of Melt Fracture and its Relation to Extrusion Behaviour'. *S.P.E. tech. Pap.*, **6** (1960)

9 METZNER, A.B., CARLEY, E.L. and PARK, I.K., 'Polymeric Melts', *Mod. Plast.*, **37**,133 (1960)

10 CORBETT, H.O., 'Extrusion Die Design', *S.P.E. Jl*, **10** 15 (1954)

11 BARTRUM, D.E. and STEBBINS, W,S., 'Anodized Aluminium Dies for PVC Extrusion', *Mod. Plast,*, **47**, 152 (1970)

12 SHELL RESEARCH LTD, Br. Pat. 862 941 (23 December 1959)
13 FISHER, E.G. and MASLEN, W.A., 'Some Aspects of Extrusion Die Design', *Br. Plast.*, **31**, 276 (1958)
14 FLATHERS, N.T. *et al.*, 'Extrusion of PVC Dry Blends with a Vacuum Hopper and Fine Screens, *Int. Plast. Engng*, **1**, 256 (1961)
15 SCHOTT, H. and KAGHAN, W.S., 'Temperature Profile of Molten Plastic Flowing in a Cylindrical Duct', *S.P.E. J1*, **20**, 139 (1964)
16 BERNHARDT, E.C., BOHRES, W.G., EDWARDS, W.M. and SQUIRES, P.H., 'Improved Instrumentation for Screw Extruders', *S.P.E. J1*, **15**, 735 (1959)
17 KESSLER, H.B., BONNER, R.N., SQUIRES, P.H. and WOLF, C.F.W., 'Rapid Response Recordings of Extrudate Temperature and Pressure — Typical Use in Evaluating Extruder Performance', *S.P.E. J1*, **16**, 267 (1960)
18 ECKMAN, R.L. and PETTIT, G.A., 'The Application of Pressure Measurement to the Extrusion Process', *Int Plast. Engng*, **1**, 179 (1961)
19 FAIRFIELD, R.M. and COX, F.J., in *Polythene*, Ed. Morgan, R. and Renfrew, A., Iliffe, London (1960), p. 467
20 TUNSTALL, H.A., 'Extruded Plastics Covering for Electric Wires and Cables', *Trans. Plast. Inst., Lond.*, **18**, 33 (1950)
21 BRITISH CELLOPHANE LTD, Br. Pat. 713 841 (26 March 1952)
22 BRITISH CELLOPHANE LTD, Br. Pat. 714 194 (23 February 1953)
23 BRITISH XYLONITE LTD, Br. Pat. 664 412 (4 January 1949)
24 A. HAGEDORN & CO. A.G., Br. Pat. 894 335 (18 August 1958)
25 HERMANN BERSTORFF MASCHINEBAU ANSTALT G.m.b.H., Br. Pat. 890 138 (12 February 1960)
26 ANON., 'Cellulose Ester Plastic Sheeting Produced by Non-Solvent Continuous Extrusion', *Mod. Plast.*, **23**, 132 (1946)
27 SCHWEIZERISCHE ISOLA-WERKE, Br. Pat. 712 367 (7 May 1952)
28 PLASTIC TEXTILE ACCESSORIES LTD., Br. Pat. 836 555 (26 October 1956), 836 556 (26 October 1956)
29 TENAPLAS LTD, unpublished notes from the Development Department, January 1946
30 JARGSTORFF, G. and JOSLIN, C., 'Dimensional Control in Extruded Tubing', *Mod. Plast.*, **26**(1), 191 (1948); BAKELITE CORPN, Br. Pat. 634 445 (16 February 1946)
31 SLAUGHTER, C.E., U.S. Pat. 2 663 904 (12 December 1950)
32 REIFENHÄUSER, H., Br. Pat. 780900 (15 July 1955)
33 S.A.S. LAVARAZIONE MATERIE PLASTICHE (L.M.P.) DI M.I. COLOMBO & CO., Br. Pat. 810 703 (1 March 1957)
35 SCHMIDT, D.J., 'Techniques in the Extrusion of High Density Polyethylene Tubing and Profiles', *S.P.E. tech. Pap.*, **6** (1960)
36 FISHER, E.G., in *Polythene*, Ed. Morgan, P. and Renfrew, A., Iliffe, London (1960), p. 483
37 PLAX CORPN, U.S. Pat. 2 443 289 (4 May 1944)
38 PLAX CORPN., U.S. Pat. 2 365 375 (23 April 1941)
39 FAIRFIELD, R.M. and COX, F.J., in *Polythene*, Ed. Morgan, P. and Renfrew, A., Iliffe, London (1960), p. 469
40 STOTT, L.L., U.S. Pat. 2 719 330 (13 March 1951)
41 THE POLYMER CORPN., U.S. Pat. 2 747 224 (6 August 1952)
42 N.V. ONDERZOEKINGSINSTITUUT RESEARCH, Br. Pat. 774 430 (24 November 1954)
43 ANON., 'Automatic Extrusion System Controls Gauge in Both Directions — Automatically', *Mod. Plast. Int.*, **4**, 62 (1974)

9

The Complete Extrusion Process

9.1 Introduction

The practical art of extrusion which consists of the manufacture of saleable products of high dimensional uniformity and general quality at economic output rates requires considerably more than the mere provision of a uniform melt of correct form at the die orifice. Regardless of the section being produced, some form of take-off or other handling system is always necessary and the quality of the finished product, once it has left the extruder, is completely determined by the regularity and smooth functioning of such mechanism. It is, therefore, impossible to consider the extrusion process seriously unless various standard types of take-off system are discussed in detail.

In general the take-off system draws the material from the extrusion die under controlled tension into and through a cooling system where the product is sized and set to required form. Further cooling takes place in the majority of applications and the finished section is presented to the wind-up, cut-off device or other handling system. Previous chapters have discussed the screw extruder mechanism in detail and have dealt with extrusion dies and sizing systems. This chapter will complete the process by discussing in detail the various methods of take-off used in extrusion operations.

9.2 Wire covering

This process is used to coat continuous lengths of wire, cable, tube, and a variety of products with a layer of thermoplastics material. The tremendous demand for covered cable in the radio and

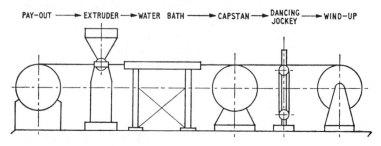

PAY—OUT ———► EXTRUDER ——►WATER BATH ————►CAPSTAN ——►DANCING JOCKEY ———►WIND—UP

Figure 9.1. Diagram of wire-covering process

electrical industries makes this one of the most important of the many extrusion processes.

The basic requirements for the complete process are shown diagrammatically in *Figure 9.1*, from which it may be seen that there are six units including the extruder in the complete arrangement.

Drums of uncovered wire are mounted on a payout stand and these drums may be free to rotate, friction braked or power driven, depending on their size and on the requirements of the system. From the payout drum the wire is led through the covering crosshead which is mounted on the extruder and which has been described in detail in Chapter 8. On leaving the crosshead, the coated wire passes through a water trough, where the hot thermoplastic covering is cooled, and then on to a capstan, which continuously hauls the wire through the whole system at a constant rate. The wire then enters a constant tension device which controls the speed of a reel-up drum on to which the covered, cooled product is finally reeled.

A modern extrusion line for high speed wire covering, in which speeds as high as 20 m/s (4000 ft/min) are not uncommon, requires a number of additional devices for control, etc., beyond the six basic units referred to above. Some of these units are therefore included in the following description.

9.2.1 PAYOUT

The payout is essentially a reel stand and on rudimentary setups the reel of uncovered wire is merely supported on centres so that it may revolve as the wire is taken off and some form of simple braking is provided to prevent the inertia of the drum from affecting the tension of the wire. In more complex systems and for fine wires, the payoff reel is driven so that a greater degree of control may be exercised on the tension of the wire throughout the system and to minimise the tension on wires of very small diameter. The payout drum may be mounted with its axis at 90° to the line or parallel to it. In the latter case the whip of the conductor as it pays out is controlled by a metal bell-shaped hood through which it is passed.

A further desirable feature on a payoff unit is the facility for mounting two reels simultaneously so that wire from a fresh reel may be introduced to the system with the minimum speed change when the preceding reel runs out. In order to effect this changeover efficiently it is necessary to have a control mechanism on the drive which enables the peripheral speed of the fresh reel to be matched to the linear speed of the outgoing wire and some method of joining the start of the new reel to the end of the old one.

9.2.2 COOLING BATH

The next unit shown in the arrangement in *Figure 9.1* is the extruder and covering crosshead, which have been described elsewhere in this work. In this connection, however, it is of interest to note that in modern wire-covering arrangements the extruder drive is coupled either electrically or hydraulically to the payout, capstan and the take-up, so that the speeds of these units increase and decrease in step.

From the extruder the wire passes into a cooling bath, or trough, the size of which varies according to the diameter or volume of the thermoplastics material to be cooled. Different covering thicknesses require different degrees of cooling and the troughs are, therefore, generally made up in sections to enable the correct length for any particular job to be assembled. The cross section of the trough depends on the size of the covered wire or cable it is to handle; 75 x 75 mm (3 x 3 inch) is a common size for ordinary insulated cores, but for large cables sizes up to 380 x 380 mm (15 x 15 in) are used.

At both ends of the cooling bath the wire passes over V-shaped weirs or through rubber seals which reduce water leaks and along the length of the bath it is sometimes held submerged by free running rollers. For some materials and thick-walled coverings the bath is divided into sections in which the cooling water is maintained at different temperatures so that gradual cooling may be effected.

The length of the cooling trough can be calculated provided the properties of the material, the conditions of extrusion—i.e. output rate, stock temperature and extrudate volume—and the temperature of the cooling water are known. Examples of such calculations have been published[1]. It is also possible to calculate the quantity of cooling water required for such a set of conditions.

With increasing extrusion speeds the length of time which the coated wire spends in the cooling water has become progressively less with the result that in recent years cooling troughs of 30 m (100 ft) or more in length have become common. Most installations are equipped for water recirculation and cooling and where several extrusion lines are in continuous operation a central recirculating and cooling system, serving all the machines, is often employed.

An important factor which affects the appearance of the insulation is the setting of the cooling bath relative to the extruder. The distance which the hot coated conductor must travel before contacting the cooling water—known as the air-gap—allows the surface of the coating to become annealed and to obtain a gloss as well as reducing strains. The gap length of this is varied according to the linear speed of the coated wire and can be as much as 1 m (3.3ft) or more on high speed installations.

As the coated wire or cable leaves the cooling trough it is usually given a mechanical wipe followed by an air-wipe to force excess water, picked up by the rapidly moving product, back into the trough and to dry the coating before it is measured.

9.2.3 MEASURING

Measurement of the length of covered product is generally carried out by means of simple linear measuring unit operated by the wire or cable passing between two or more rollers or wheels. Refinements include pre-set length indication and alarm signals and adjustments for various diameters of coating.

9.2.4 TESTING

It is common practice to include test apparatus in the extrusion line so that immediate correction of faults can be carried out. The diameter of the coated conductor can be measured mechanically or electrically, the latter method by sensing electronically the capacitance of the insulant at a particular point. Any variation in diameter is noted as an electrical impulse which may be fed into the capstan drive to adjust the speed and effect correction. Lack of conductor concentricity can also be indicated by an extension of this capacitance principle but no satisfactory system of automatic eccentricity correction has been developed so far although there have been several attempts to do so.

Spark testing for insulation faults is provided by passage of the coated wire through a high voltage field or by sliding contact with a metal electrode at high potential. Some testers also mark the cable where the fault occurs so that the faulty section can be cut out or repaired later, the method of marking depending on the design of the test equipment but frequently the destructive effect of the spark is sufficient.

9.2.5 CONDUCTOR PREHEATING

When PVC, polyethylene or other thermoplastic is extruded over a cold conductor, the heat is rapidly conducted away so that the thermoplastic becomes prematurely chilled and unable to reach a strain-free state. An insulated wire or cable produced under such conditions would have poor physical properties and the insulant would tend to shrink back on reheating and crack at low temperatures.

Preheating of the conductor just prior to its entry to the crosshead overcomes these difficulties and is carried out as normal practice in the majority of wire and cable extrusion operations. The conductor must be heated to a temperature just below the processing temperature of the thermoplastic and this is most conveniently achieved by passing an electric current through a short section of the conductor so that heat is generated by ohmic resistance.

Contact with the bare conductor is made by means of one or more pairs of rollers which are connected to a step-down transformer giving a low and safe voltage supply.

For very large cables, which would require very high currents and massive equipment for adequate resistance preheating of the conductor, gas and high frequency induction heaters have been used.

9.2.6 CAPSTAN

The capstan, as its name implies, consists of a large winding drum—or two drums—around which the wire is passed four or five times before being led to the take-up or wind-up stand. The diameter of the drums must be suited to the type and dimensions of the wire being produced and may vary from 300 mm (12 in) for very fine wires up to 1.5 m (5 ft) or more. The drive for the capstan is capable of variable speed adjustments with precise control which may be linked to the speeds of the extruder, payout and take-up. It is usually an advantage to provide additional cooling to the wire by running the capstan drums in a water trough and if this plan is adopted it is essential to ensure that the testing referred to in Section 9.2.4 is carried out after the wire has been completely dried.

For heavy cables of large diameter a capstan pulling system is not practical because of the distortion which would occur in passing the product around the drums. For cables of such dimensions a caterpillar traction unit is employed. This consists of two or more driven belts adjusted to grip the cable and by their movement to pull it through the system. The tractive force available with these caterpillar units is very high and several sizes are available to suit the cables with which they are to be used.

9.2.7 TAKE-UP

The final unit in the wire covering take-off is the take-up stand which is most of its features is very similar to the payout. The unit, ideally, caters for two reels and is is necessarily power-driven. A further refinement of this unit is a laying device which guides the wire on to the reel in such a way that it is neatly and regularly laid into position and does not bunch in one spot. It should be possible on this unit also to transfer the incoming wire quickly and automatically from a full reel to an empty one with the minimum retardation of the system and the laying device must also transfer itself automatically to the empty drum. An automatic cut-off is also necessary and the rotation of the full drum must be stopped automatically so that it can be unloaded and replaced without stopping the operation.

Because of the change in diameter of the coil of finished wire on the wind-up drum as production proceeds there must be some mechanism interposed between the capstan and the wind-up which will automatically control the tension on the wire by progressively reducing the wind-up speed. Such a mechanism must also accommodate the erratic changes in tension which occur when drums are changed, etc. In simple installations this can be achieved by using a slack belt or a slipping clutch to drive the windup from the capstan but in more elaborate setups the 'dancing jockey' system is used. One design of this unit consists of two pulley sheaves, one of which is held in a fixed position whilst the other can move freely to enable the gap between them to be varied. The tension on the wire is balanced by an adjustable weight which holds the two pulleys apart. Any change in tension of the wire results in a change in the position of the moving sheave and this movement is connected electrically to the wind-up drive which compensates for the increase or decrease by accelerating or decelerating respectively. Other designs include mounting the pulleys on pivoted arms which open and close against a predetermined tension.

The take-off equipment outlines above can be adapted for use with the majority of covering applications. A few modifications are required as the diameter of the wire or cable increases. The 'dancing jockey' arrangement becomes less important as the diameter of cable goes up since tension changes are no longer critical from a breaking stress point of view. Furthermore, since the minimum bending radius of large cable is very great the system necessary to control such material would be prohibitive in size. The problem of handling this becomes more closely allied to that of pipe take-off and the speeds are correspondingly lower.

9.2.8 SPECIAL EQUIPMENT

Additional equipment is also available for various other operations and this includes roll stands for supporting heavy cables between the various stages of the take-off and so that the equipment need not be laid out in a straight line in the event of space limitation. There are also continuous printing devices for marking the cable and equipment for flame polishing. For special high temperature work a crosslinked polymer insulant is sometimes required and to provide for this electron accelerators are occasionally incorporated into the extrusion line to irradiate the cable just prior to reeling. An alternative method uses chemical crosslinking.

An extrusion line for wire covering with a resin which requires thermal curing has been described by Rhodes and Black[2]. The wire is fed vertically downwards through the extrusion die into and through a curing tunnel. It is then cooled, first by water spray and then by passage through a cooling bath, before it is finally wound onto a capstan and take-up unit of normal design.

Accurate measurement of the diameter of insulated wires and the maintenance of the prescribed diameter within close limits is an important aspect of the economics of wire and cable production which was touched on briefly in Section 9.2.4. At linear speeds of 10 m/s (2000 ft/min) or more a very small oversizing can quickly use up a large amount of excess material for which the producer will not be paid. A photoelectric gauge accurate to $\pm$ 6.3 μ m (0.00025in) is described by Zimmermann[3]. This apparatus corrects variations from a set dimension by adjusting the payout and capstan speeds, the extruder speed being held constant. The correct functioning of this device calls for very accurately controlled synchronised drives and special controls, some of which have been described[4,5].

A later instrument developed for the same purpose uses, in principle, a narrow beam of light to sweep continually across the cable at constant speed. The diameter of the cable is given by measurement of the time interval during which the light beam is interrupted. Such an instrument is obviously not affected by variations in intensity of light source or of photocell sensitivity and amplifier gain[6].

9.3   Sheet and flat film

There are two methods of extruding sheet material and for convenience these two processes can be referred to as the direct and the indirect methods.

The direct method consists of extruding a sheet or film from a flat die, similar to the types described in Section 8.4.4. and thence through a roll take-off system to a wind-up or cut-off mechanism.

The indirect method utilises a tube die of the straight or the crosshead type and the extruded tube is continuously slit and opened to a flat form or blown in a manner which is described later and slit subsequently.

9.3.1   DIRECT FILM EXTRUSION

The take-off system used for the production of film products by the direct method is usually built into a compact self-contained unit. *Figure 9.2a* is a diagram showing the layout of one design of such a unit. The film is extruded vertically downwards from the die and enters a water bath, the temperature of which is rigidly controlled[7]. In the scheme shown in *Figure 9.2a* the film changes its direction by passing around an idler roll, which runs in the water bath, and then passes through two driven nip rollers. Between the nip rolls the film is cut to the required width by means of rotating trimming knives and passes finally to the wind-up reel. A further layout for a water-quench film line as suggested by Haine and Robinson[8] is shown in *Figure 9.2b*.

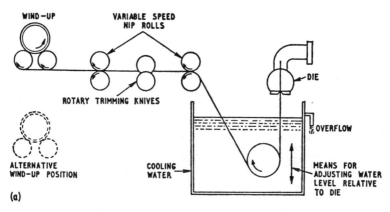

Figure 9.2a. Diagram of water quench process for flat film

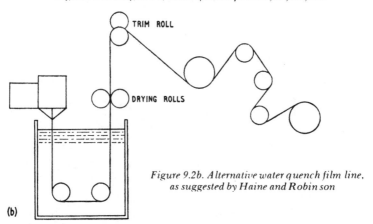

Figure 9.2b. Alternative water quench film line, as suggested by Haine and Robinson

The final wind-up spindle can be directly driven through a slipping clutch mechanism to ensure constant tension on the web or, alternatively, a peripheral friction drive may be used which maintains a constant linear take-up speed regardless of the reel diameter. Specially designed constant tension electric drives are also available for this application. Twin wind-up stations enable a fresh reel to be started and a full one removed without stopping the process.

The rolls on a film take-up unit such as that described above are usually of polished steel and often have their surfaces chromium-plated, since imperfections on the rolls are transferred to the sheet as a recurring pattern. The nip rolls usually consist of a steel roll and a rubber-surfaced pressure roll which has been ground to a smooth parallel finish.

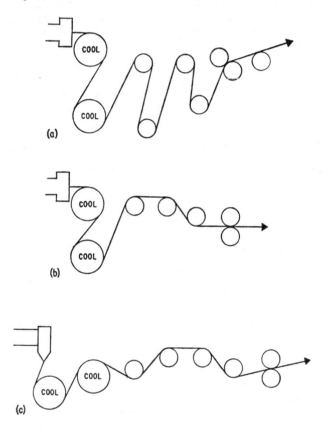

Figures 9.3a, b, c. Schematic arrangement of various chill roll film lines

An alternative method for the production of films, particularly the medium density polyethylenes and polypropylene is to pass the extrudate over chilled rolls rather than directly into water as described in the preceding example. The chill roll system is said to give higher strength properties, which, moreover, can be con-trolled by adjusting the extrusion conditions, and higher clarity when operated with suitable polymers. The water quench method gives a stiffer, glossier film with a lower surface coefficient of friction, but it is difficult to avoid marking the film due to disturbances of the quench bath surface. Detailed descriptions of both methods of working have been published[8-10].

Whichever method is used, i.e. the water quench or the chill roll system, the length of the air gap, or hot stretch distance as it is sometimes called, between the die outlet and the cooling means has an important bearing on the properties of the final product. In the water quench method this gap should be adjustable from about 6 mm (¼ in) up to 150 mm (6 in) or more whereas with the chill roll system air-gaps of up to 75 mm (3 in) are used. Obviously the air-gap length will be determined to a large extent by the linear speed of output so that higher production rates generally require longer gaps and vice versa.

A frequent trouble which occurs in the extrusion of film is marring of the surface of the product due to vibration of the equipment. When the water quench method is used even small disturbances of the water surface will affect the appearance of the film and in the chill roll system it is particularly important that the rolls should be dynamically balanced.

Various layouts for chill roll film extrusion lines are shown diagrammatically in *Figures 9.3a, b* and *c*. The rolls are water-cooled and designed to give maximum and uniform heat transfer and although two rolls only are shown in the illustrations it is not uncommon for three such rolls to be used in high speed installations.

9.3.1.1 Orientation equipment

By stretching the film in one or both directions under correct conditions of temperature the molecules become lined up and the product is said to be orientated. The high strength of man-made fibres such as nylon and polypropylene monofilaments, etc., is largely due to an orientation process.

Correctly biaxially orientated film offers several improved properties which are of interest to converters. Such products are usually tougher than the unstretched material so that a thinner film can often be used, are 'crisper', and usually show higher clarity. Polystyrene film is biaxially orientated, as are the polyethylenes, polypropylene and polyester films, to give materials of high strength and great value to the packaging industry. Films biaxially orientated under correct conditions can also be used for 'shrink' wrapping.

The problem of orientating the material in the direction of extrusion is comparatively simple as the film can be drawn during take-off. Stretching the film in the transverse direction is a more complex matter, however, requiring costly equipment and considerable skill. The equipment which is frequently used for this process is the tentering frame or stenter which has been adapted from similar machines used in the textile industry.

In one method the film or sheet is extruded onto a polished roll where it is surface cooled. It then passed to heated rolls where the film is brought up to the correct orientation temperatures, which varies with the characteristics of the material, and then enters the tentering frame. Orientation of the film along its length is accomplished by running the take-off rolls at the exit end of the tentering frame at a correctly adjusted excess speed. At the same time a series of grippers attached to a pair of endless chains hold the edges of the film and travelling on diverging rails cause transverse stretching as the product moves through the frame. The whole unit is enclosed in tunnel ovens which control the temperature in each section at levels suitable for stretching and for subsequent cooling.

An alternative method of transverse orientation is to stretch the film between two pairs of diverging belts by which it is held. The above and other methods and detail refinements are described in the rather complex patent literature[11].

9.3.1.2 Film tape yarn

An important application recently developed for the use of polypropylene and high density polyethylene films is in the production of narrow split mono-axially orientated film tapes in widths from about 1.5 mm upwards and starting thickness of 100 μm, or thereabouts, for weaving into primary backing fabric for carpets and into sacks, bale wraps, etc. as a replacement for woven jute materials. Such tape yarns are lighter and stronger than natural fibre materials, do not rot and have superior resistance to chemical attack.

If the film tape is subject to maximum orientation, i.e. a stretch ratio of 1 : 10 or more, the product can be fibrillated by various mechanical and other means into monofilaments for subsequent processing into continuous twisted yarns and staple fibre[50-52]. In an extension of this process Smith & Nephew have produced net structures by the introduction of biaxial orientation to a polymer film which has been previously melt-embossed in a pre-determined diamond or other pattern[53].

In the production of film tape yarn the film, of thickness suited to the particular application, which may be produced flat by either of the methods described in Section 9.3.1 or tubular as in Section 9.4, is slit into a number of tapes of required width and passed to the first godet of the orientation equipment and thence through a controlled temperature oven to the second godet. Controlled orientation is thus imparted to the tapes which are then usually—but not always—relaxed or annealed in a second oven and thence pass to a third godet roll and finally to a multi spindle wind-up.

In general the degree of orientation—i.e. the stretch ratio—the stretch and annealing temperatures are dependent on the particular application, the grade of material used and the method of film production. Stretch ratios vary from 1 : 5 to 1 : 7, for tape yarn used in carpet backing and sack production and up to 1 : 14 for fibrillation. Stretch temperatures can vary from 120 to 170 °C although the optimum is usually in the region of the softening point of the material used. Annealing temperatures are usually similar to stretch temperatures or slightly higher.

There are several variations possible to the general technique outlined above for the manufacture of these products. In one such variation the complete unit is split into two stages whereby the film is separately produced and wound and is subject to splitting and orientation as a separate operation. By this means both units can be arranged to operate at maximum efficiency. The capital cost is, however, higher and the film properties may deteriorate or change during storage.

Another variation is known as hot roll stretching, whereby the film is stretched in its full width on preheated rolls giving a roll of stretched film which can be used as a warp beam for slitting directly in the loom. This method, avoiding as it does the conventional orientation equipment, the controlled-temperature ovens and the tape winding equipment, makes for a very compact unit. It is suitable only for a limited number of applications however.

9.3.2 SHEET EXTRUSION

It is often not possible to reel thick sheet material nor is it usually advisable to pass the thick materials through a water bath. The type of take-off system used for the production of polystyrene sheet, for example, is shown diagrammatically in *Figure 9.4*. On leaving the die the sheet passes round a triple roll system in which the rolls are heated and maintained at constant temperatures.

After leaving the initial polishing rolls the sheet is supported on a belt or roller conveyor and is led into one or two pairs of nip rolls with a set of edge trimming knives introduced between the roll stations. For sheet materials which cannot be conveniently reeled it is necessary to cut the web into lengths and for this purpose a guillotine is provided at the end of the take-off.

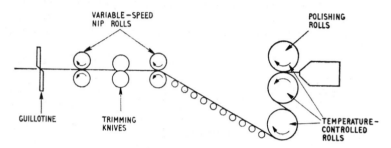

Figure 9.4. Diagram of process for extruded sheet

The arrangement for producing sheet by the indirect method consists merely of a large tube-die of small wall thickness, which produces a complete tube of material as described in Section 8.4.4. Immediately the tube leaves the die it is slit by a stationary cutter and is opened out gradually over a series of specially shaped guides. When the sheet has been flattened it is taken round one or more heated double-roll assemblies and is finally reeled or cut into lengths as required.

9.3.2.1 Polishing rolls

Although at one time it was common practice for two-roll polishing stacks to be used in sheet extrusion, three-roll vertical arrangements as shown in *Figure 9.4* are now almost universal. It is usual also for the material from the die to be fed into the upper nip and to travel down the stack as shown also in *Figure 9.4*. Alternatively the extrudate can be fed into the lower nip and travel upwards, the reported advantage of this latter method being that the polished surface is then face up and cannot be damaged by contact with subsequent conveying equipment.

The polishing rolls are usually hard chrome plated and various diameters up to about 300 mm (12 in) are used. In all cases the rolls are provided with individual temperature control, usually by means of oil, and must be designed internally to give uniform heat transfer over their entire surface. The speed of the rolls must be controlled to correspond with the output of the extruder or fractionally faster in order to compensate for the slight swelling of the extrudate as it leaves the die. The polishing unit is in no sense a calender and cannot therefore reduce the thickness of the sheet nor is it a haul-off unit to introduce draw-down. The function of the unit is to impart a good surface to the product without warpage and to reduce its temperature to a level at which it can be handled without distortion.

Figure 9.5 1.8m wide sheet line showing polishing stack using 400mm diameter rolls (Courtesy N.R.M. Corpn, Akron, Ohio

In the case of some high impact polystyrenes it is difficult to obtain a high polish with the rolls alone as these materials tend to be naturally dull owing to the presence of the modifying polymer. To overcome this a thin film of biaxially orientated unmodified polystyrene is sometimes laminated to the sheet as it enters the polishing unit and provision for supporting the roll of film above the polishing roll stack is made on most units (*Figure 9.5*).

In operation the temperature of the top two polishing rolls should be as high as possible without sticking and the bottom roll should be just cool enough to prevent distortion of the sheet during its passage to the next stage in the take-off unit.

9.3.2.2   Conveyor and pulling rolls

From the polishing rolls the sheet usually passes along a conveyor which may consist of free-running rollers or wheels, or a driven belt conveyor may be used. In order to assist in the cooling of the product air blowers may be provided at this point both above and below the sheet and for this reason the roller system of conveying has definite advantages over the belt type, which obviously prevents cooling of the lower surface. The conveyor may be up to 3.5m (twelve feet) in length and is inclined upwards at a suitable angle to allow the sheet to enter the pulling roll nip.

The pulling rolls are rubber- or neoprene-coated and their speed is adjusted to be slightly less than that of the polishing rolls to allow for the shrinkage which occurs in the sheet as it cools.

9.3.2.3   Trimming, cutting and stacking

Edge trimming, if this is necessary,—with sheet intended for vacuum forming this is not always so—may take place just prior to the nip rolls or a second set of nip rolls may be provided with trimming facilities between the two sets as previously referred to. On thin sheets razor blade cutters may be used but rotary shears are necessary with thicker products.

9.3.2.4   Final handling equipment

After passing through the nip rolls the sheet is either coiled on conventional equipment or if it is too thick for such treatment it passes to the guillotine where it is cut into accurate lengths and automatically stacked on a trolley or stillage. The guillotine is usually electropneumatically operated and actuated by the sheet itself through the medium of correctly placed microswitches of photoelectric equipment. In order to protect the surface of the sheet, tissue paper is sometimes fed with it throught the nip rolls and the guillotine to interleave the stacked sheets.

9.3.2.5   Additional equipment

In addition to the above basic equipment continuous thickness gauges are sometimes built into the take-off unit. These can be of the mechanical, optical or beta ray type. Mechanical measuring units to be of use must be very accurately made and rigidly placed and are generally not satisfactory. Beta ray equipment is very satisfactory but costly. Optical units are said to be satisfactory and much less costly.

During passage through the take-off equipment most thermoplastic sheets will acquire a static charge which will attract dust and spoil the appearance of the product. Static eliminators of one form or another are, therefore, frequently built into the equipment. The sheet can also be passed through a detergent bath or sprayed with detergent solution.

9.3.3 SPECIAL SHEET PROCESSES AND TECHNIQUES

9.3.3.1 Corrugating plant

The development of corrugated rigid PVC sheet for use as roofing and in other applications has been extremely rapid during the past few years and the development of plant for its manufacture has been correspondingly rapid. Apart from the extruder and the die, which have been described in other chapters, the process can be considered under two categories; firstly the production of continuous sheet with transverse corrugation and secondly the production of standard sheets longitudinally corrugated. Both products can be produced from the same extruder and flat sheet die provided the necessary corrugating equipment is available.

Transverse corrugations. Several methods have been used successfully for the forming of transverse corrugations in extruded sheet as a continuous process. In most cases the sheet as it emerges from the die is passed through a two- or three-roll polishing unit in the normal manner and is then reheated. The flat sheet so produced is fed between two sets of driven chains on which are fitted a number of freely rotating rolls so placed that the two sets mesh with each other and in so doing form corrugations in the sheet. The diameter and positioning of the forming rollers can be changed to give different profiles.

A further method uses a profiled or corrugated belt and six or so water-cooled rolls so fixed above the belt that they can only move vertically up or down. As the belt with the sheet moves along horizontally the rolls ride over the belt profile forming the sheet as they do so. One type of transverse corrugating equipment is shown in operation in *Figure 9.6*.

Figure 9.6 One form of transverse corrugating unit in operation. (Courtesy Klein-wefers Sohen, Krefeld, Germany)

Figure 9.7. An extrusion line in operation producing longitudinally corrugated sheet. (Courtesy Anger Gebruder G.m.b.H., Munich)

Longitudinal corrugation. Although there have been several attempts to produce longitudinally corrugated sheet direct from a suitable profiled extrusion die—and in fact, this does not seem to present any greater difficulty than extruding the flat sheet—it is generally conceded that post-forming is the more economic method. The sheet as it leaves the polishing unit is reheated as before and pre-formed by means of corrugating wheels. It is then passed to a corrugating die which determines the final contour. Subsequent operations are guillotining and stacking as described for flat sheet. Edge trimming is also necessary so that the finished sheets confirm to established size standards[39].

One form of longitudinal corrugation unit is shown in operation in *Figure 9.7*.

9.3.3.2   Floor tile plant

A plant for the continuous production of floor tiles by extrusion has been developed as an alternative to the use of heavy calenders and internal mixers. If successful such equipment would show a considerable saving in capital investment and in labour costs. The possible output of the plant would, of course be considerably lower but this is not necessarily a disadvantage as several units could be used to give greater product flexibility.

The plant is, in fact, a special extrusion line which has been designed to produce heavy-gauge plasticised and filled vinyl sheeting. Instead of the normal polishing roll unit a more robust three-roll calender is used. As the flooring must be stress-free the sheet is annealed by passage through a temperature-controlled tunnel oven after which it is cooled on chilled rolls before coiling or cutting into finished tiles. Very accurate take-off speed control is essential in order to avoid setting up further stress in the product after it leaves the calender.

9.3.4   SPECIAL SHEET MATERIALS

Certain thermoplastics materials and products do not behave well in the typical sheet extrusion lines which have been described so far and various modifications, either in equipment or in operational techniques, are necessary to obtain saleable products.

9.3.4.1   Polycarbonates

Instead of the normal three-roll polishing unit a straight two-roll calender is preferred for these materials. In operation the rolls function rather as a conventional calender than as a mere polishing unit, thus effecting shear to the material as well as sizing it.

9.3.4.2   Methacrylates

The sensitive temperature-viscosity relationship of these materials causes the extrusion of sheet to become a highly critical operation calling for accurate processing conditions. A two-roll polishing unit—not a calender—is advised instead of the three-roll stack and it has been suggested that an upward extrusion and take-off system improves optical clarity and reduces residual stress[40].

9.3.4.3   Sheet in foamed plastics

The manufacture of foamed sheet by the extrusion process can be carried out in various ways depending on the material. For thick sheets in expanded polystyrene for example, the standard flat die may be used although the surface finish may be poor owing to residual expansion which continues after the material has left the die. The second and more popular method which is used for the production of thinner sheets and of lower density in the same material follows the tubular film technique which is described in a later section. Using a circular die a blown sheet is developed, fed to take-off rolls, edge-trimmed and the resulting two sheets reeled up[12].

9.4   Tubular film

One of the more important special extrusion processes is the manufacture of layflat tube or tubular film. The type of die used is described in Chapter 8 and *Figure 9.8* shows a schematic layout of one form of take-off for this process.

It is seen to consist of a framework which supports a pair of nip rolls vertically above the die and which also has mounted on it a take-up reel on to which the flattened tube is wound. One of the nip rolls is usually steel with a ground and polished surface and the other has a ground rubber surface. The rolls are geared together in the same way as the rolls of a domestic wringer and can be spring or hydraulically loaded one against the other. A variable-speed drive with a wide speed range is coupled to the nip rolls and the wind-up spindle is also connected to the same drive by some suitable constant tension device. On some simple take-offs the reel of material merely rests on one of the nip rolls and is driven by friction at the same peripheral speed as the flat film, thus obviating the necessity for a constant-tension device.

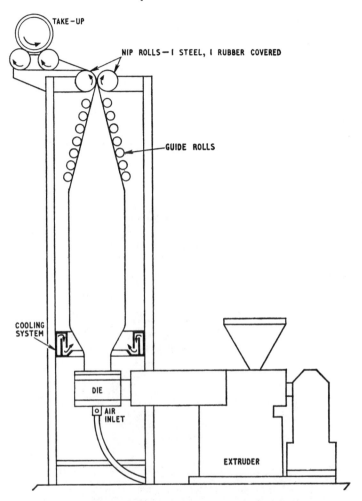

Figure 9.8. Diagram showing one process for tubular film extrusion

As the molten tube emerges from the die it is inflated over the constant air mandrel which is introduced, at the commencement of the process, through the stopcock at the bottom of the die. Some form of cooling device is placed above the die through which the inflated tube passes and a cooling medium is projected on to the tube to set it to the required diameter[13-16]. The tube then passes between two sets of guiding rolls which flatten the bubble and lead it to the nip rolls. Shaped hardboard surfaces or wire mesh stretched on frames may be used in the place of the guiding rolls. As the tube passes between the nip rolls it is pressed completely flat and the air bubble in the tube above the die is unable to escape.

Twin wind-up stations are provided on most tubular film take-offs to enable the process to continue whilst a full reel is removed. Provision is also made on some arrangements to trim the edges of the flattened tube and so to produce two thicknesses of flat film simultaneously. The complete tubular film equipment is shown in *Figure 9.9* where many of these features can be clearly seen.

Figure 9.9. Complete tubular film plant in operation. (Courtesy N.R.M. Corpn, Akron, Ohio)

A further development consists of the provision of a gussetting device whereby the tubular film is given two tucks or folds which facilitate bag manufacture and use. This device may be in the form of a pair of forming members arranged between the guiding rolls so that the tubular film is continuously folded inwards on opposite sides[17,18].

The diameter of the blown film is usually between one and three times the diameter of the die orifice, so that one die can often be used to produce a wide range of film widths. As well as blowing the tubular film and thus reducing its wall thickness by lateral stretching, the rate of take-off by the nip rolls is adjusted to give a longitudinal draw-down and in this way the degree of orientation and thus the relative tensile strength of the film parallel to and perpendicular to its length is adjusted.

Some tubular film take-off systems are arranged so that the film is taken off vertically downwards, the extruder being supported on a steel tower or on an upper floor. This system has the advantage of easy handling of large rolls of film since the finished product is reeled at floor level and does not require additional guide rolls to bring the product down from the nip rolls. It is, however, necessary to carry raw material to the machine in an elevated position and there are sometimes difficulties due to the cooling medium rising to chill the die orifice.

A further take-off system involves horizontal extrusion and this process is in many respects a modified tube take-off. Such a system has the advantage of being easier to construct since neither the extruder nor the wind-up have to be supported above the ground. There are, however, disadvantages attached to this system in connection with distortion in the film under some circumstances due to gravitation effects.

In a study of extrusion variables Clegg and Huck have published information on the position of the 'freeze' line on the bubble, bubble cooling and the overall shape of the bubble[19]. A more general survey has been made by Grant[20], and the effects of different blow-up ratios have been described by Pilaro *et al*[21].

The output-limiting factor in all tubular film processes is the linear rate at which the product can be cooled sufficiently to avoid 'blocking', or sticking together of the two sides, due to pressure from the nip rolls. As this linear cooling rate is roughly similar irrespective of the diameter of the tubular film being produced, the process becomes less economical in output in terms of kilogrammes per hour as the size is reduced. Many firms, therefore, use a number of small machines of perhaps 40mm (1½in) screw diameter for producing the smaller sizes whilst reserving larger machines for the larger sizes.

In order to overcome this difficulty there have been many attempts to produce a number of small tubular films simultaneously from one large or medium-size machine, and this technique has been developed so that it has now become relatively common practice. The Oerlikon equipment is one example of a commercially available unit for the production of four tubular films simultaneously from one extruder.

9.4.1 SPECIAL EQUIPMENT AND TECHNIQUES

The tubular film process is a somewhat critical operation and in consequence there have been a number of ingenious developments which aim at simplifying the technique or at improving the product without increasing the difficulties.

9.4.1.1 *Rotary dies and bubble cooling systems*

Tubular film produced under normal production conditions often contains longitudinal thin or thick narrow bands which cannot be rectified by die concentricity adjustment. These bands, which are due to localised flow irregularities in the die system, and may be of the order of 25 μm (0.001 in) thickness variation, do not affect the usefulness of the film as a packaging medium and are usually undetectable to the eye. In a tightly wound roll of the product, however, they become apparent with the building up of successive layers to form a hoop in the roll, and cause cold stretching and distortion of the outer layers.

This difficulty is normally overcome by rotating or oscillating the die or the take-off so that the irregular portion is distributed around the circumference of the tube, and thus does not build up in one place on the roll. A rotating complete extruder in which this purpose is achieved is also described in Chapter 5. Rotating of the cooling system only is also used, particularly on the larger sizes, but this method is less effective.

As discussed above, it will be seen that an extremely important unit in tubular film production equipment, is the bubble cooling system, one form of which is schematically shown in *Figure 9.8*, and which consists of an air ring arranged to project low pressure, high volume cooling air onto the exterior of the bubble as it emerges from the die. Obviously it is important that the volume of this cooling air be as large as possible since the rate at which the tubular film can be cooled determines the rate of output at which it can finally be produced. However, it is also of great importance that the volume and the temperature of the cooling air presented to the moving bubble should be equal at all points around its circumference since any irregularity in this respect would result in irregular draw-down and consequent variations in film thickness.

Much ingenuity has therefore gone into the development of labyrinthine cooling rings which aim at providing the maximum volume of cooling air to the moving bubble with the maximum circumferential uniformity. Most makers of tubular film equipment have their own developed and sometimes patented cooling ring systems and some makers recommend the use of two of more such rings in series.

In addition to the external cooling rings referring to the above, there have been several attempts to provide internal cooling of the bubble also. When used successfully, internal cooling systems are stated to give improved haze, gloss and strength properties to the product as well as increasing the rate of output. Several such systems have been described[41-44].

More recent work on bubble cooling in the tubular film process has shown that water cooling systems can also be used to produce better quality products at higher output rates. In these systems the tubular film is passed downwards into a water bath, at least three methods are stated to be in current use[45].

9.4.1.2 Annealing

In order to improve the optical properties of polyethylene tubular film an annealing chamber has been described[22] which consists of a tube somewhat larger in bore than the die orifice which is interposed between the die and the cooling device. The freezing of the tube is thus delayed so that its surface can stablise.

9.4.1.3 Static eliminators

As mentioned previously, the production of film and sheet products, whether by the flat or the tubular system, usually gives trouble due to the build up of static charges. Many tubular film production lines, therefore, include antistatic devices of one form or another. Adequate reviews of the various methods used have been published elsewhere[23,24].

9.4.1.4   Surface treatment

Most thermoplastics films, in order to be of use to the packaging industry, have to be printed. Many materials such as PVC, polystyrene and the cellulosics, can be printed without difficulty if suitable inks are used. The polyolefins, on the other hand, are not receptive to printing inks, and therefore require special treatment before printing can be successfully applied. There are three general methods available, all of which aim at oxidising or otherwise modifying the film surface.

Chemical, in which the surface of the film is subjected to powerful oxidising agents such as chromic acid or ozone, or is chlorinated by passage through chlorine gas. There have been several variations of these techniques[25],[26].

Heat treatment, in which one serface of the product is subjected to intense heat for a very short period whilst the other surface is maintained at a low temperature to prevent melting. The heating medium may be hot gases, radiant heat, or direct flame[27-29].

Electron bombardment. In this method the film is passed through a high voltage field which is accompanied by corona discharge and various methods of applying the principle have been developed[30],[31]. In general, the electron bombardment method of surface treatment is the most convenient system and is widely used.

9.4.1.5   Orientation

Although the orientation of thermoplastics films has been dealt with in the section dealing with flat film production (Section 9.3.1.1) it is obvious that this technique is also of importance in connection with tubular films.

Owing to their method of production, all blown tubular films are orientated to a greater or lesser extent, depending on processing conditions, and systems have been devised for increasing the degree or orentation. Low melt temperatures are used, for example, as are higher than normal blow-up ratios and take-off speeds.

Although the methods referred to above do have a noticeable effect on the degree of orientation, it is difficult to obtain a balanced product by such means and to ensure reproducibility. Considerable thought has, therefore, been given to the development of better systems for the orientation of tubular film, and the results of some of this work have been published[11].

9.4.1.6 High density polyethylene film

A recently developed important application for the tubular film process is in the manufacture of tissue-paper-like films in high density and particularly in high-molecular-weight high-density polyethylenes. These films, which as their name implies possess a great similarity to tissue paper in feel, appearance and crispness, are now much used in retail stores for food and other wrapping and for conversion into bags, etc., where they are preferred for their much greater strength properties and resistance to moisture, etc., compared with the paper products they are replacing.

Tissue film is produced on substantially conventional tubular film plant as described previously in this section but with increased blow-up ratio and draw-off speeds and considerable attention to process detail as follows:

The extruder must produce a thermally homogeneous melt at the required throughput rate at as low a melt temperature as possible. The screw therefore should be deeply cut, to avoid high shear, but with a short intensive barrier-type mixing zone as described previously.

The die should, ideally, be of the centre (bottom) feed high pressure spiral mandrel type with a die gap thickness of the order of 0.75-1mm (0.030-0.040in). Means of changing the die gap thickness must be provided. Blow-up ratios of at least 3 : 1 are required and these may go up as high as 8 : 1 in some circumstances. These high blow ratios are necessary in order to reduce the tendency towards fibrillation which would arise if excessively high drawdown in the machine direction were used to obtain the required low thickness dimension.

The cooling ring should be of the highest possible efficiency and provided with a screw-operated shutter so that the amount of cooling air and thus the bubble temperature and also the position of the freeze line, which are both very critical in this operation, can be controlled by the operator. The cooling ring must be rigidly mounted and concentric with the die orifice.

The collapsing means should extend almost down to the freeze line of the emerging film and up into the nip of the take-off rolls and the height of the nip rolls above the die face should be adjustable between about 1.5 and 2.1 m (5 and 7 ft)—compared with 6 m (20 ft) or thereabouts for low density material. If possible the nip roll assembly—or the complete extruder—should be arranged to rotate or oscillate about the die centre line to spread the inevitable small thickness differences as previously discussed. Because of the greater rigidity of the cooled film, compared with an equivalent low density polyethylene product, there is increased difficulty in avoiding wrinkles in the collapsing and wind-up system. It is essential therefore that the tubular tissue film be collapsed at a comparatively high temperature to reduce the effects of this rigidity, and temperatures of the order of 50-75°C are recommended. In order to assist in this requirement heated collapsing means are sometimes used.

A photograph of an extrusion plant producing high molecular weight, high density polyethylene film is given in *Figure 9.10*

Figure 9.10. Tissue film production in high molecular weight high density polyethylene using rotating take-off system. (Courtesy A. Reifenhauser K.G., Troisdorf)

9.5 Extrusion coating

Equipment for extrusion coating must be capable of handling an extensive range of supporting materials or substrates and at the same time apply coatings in a variety of thermoplastics. The supporting materials range from cellulose film to wire mesh and include paper, cloth, metal foils and glass fibre webs. The majority of coating is carried out in polyethylene, but other thermoplastics are also used. Coating equipment up to 1.5 m (60 in) in width is common and greater widths up to 3 m (120 in) are also available and operating speeds of up to 12.5 m/s (2500 ft/min) are possible under correct conditions.

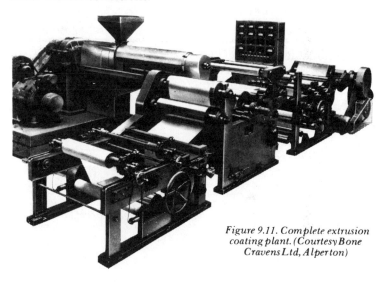

Figure 9.11. Complete extrusion coating plant. (Courtesy Bone Cravens Ltd, Alperton)

A complete extrusion coating line consists of three basic units, i.e. the unwind station, the coating and laminating unit, and the wind-up unit. For convenience of description, these units will be discussed separately, and it will be assumed that the supporting material is paper and the thermoplastic is polyethylene. *Figure 9.11* shows the complete equipment as interpreted by one well-known manufacturer.

9.5.1 THE UNWIND STATION

To ensure continuous operation, it is essential for both the unwind unit and the wind-up to have roll changing facilities without interrupting the operation of the equipment. For this reason there are two positions for substrate and finished product reels. As it leaves its reel the paper substrate is passed over several supporting rolls, to eliminate uneven stretch from previous handling, and is then passed through a preheating zone which may be a heated roll or a batch of radiant heaters. In some setups the substrate is also passed over a batch of powerful lights so that any imperfections can be detected before it enters the coating unit.

9.5.2 THE COATING UNIT

The dried paper substrate is led from the heating zone into the nip between a rubber-covered pressure roll and a water-cooled polished metal chill roll which are positioned immediately beneath the lips of a film extrusion die of suitable width. Typical sizes for these two rolls are pressure roll 230mm (9in), chill roll 400mm (16in).

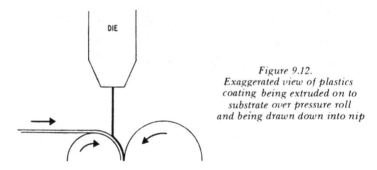

Figure 9.12.
Exaggerated view of plastics coating being extruded on to substrate over pressure roll and being drawn down into nip

The extruder temperatures are adjusted to give a very fluid melt and a curtain of molten polyethylene is extruded downwards into the nip to contact the substrate. The speed of rotation of the rolls is adjusted to draw down the polyethylene thickness to the required dimension and coating weight and the effects of pressure and cooling, and the oxidation of the high temperature polyethylene surface, causes the resin to adhere the paper (*Figure 9.12*). However, the use of very high temperatures and the reliance on consequent oxidation effects to promote adhesion to the substrate at high linear speeds has sometimes given taste and odour problems in the finished product and other adhesion promotion systems have

been developed. The use of chemical primers such as organic tita-
nates has proved successful but relatively slow. High intensity
corona discharge, which is thought to operate by carrying some of
the generated ozone into the polyethylene/substrate nip, has also
proved successful.

9.5.3 TAKE-UP UNIT

From the coating unit the product passes through adjustable edge
trimming knives to the wind-up unit which is usually of the turn-
over stand type for core changing.

The edge trimming often presents a problem because the scrap
so produced is contaminated with paper, and cannot therefore be
re-used. The technique of overcoating, i.e. applying a coating
wider than the substrate, overcomes this problem.

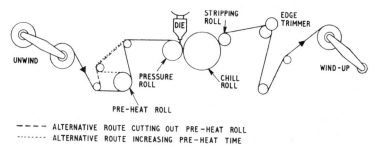

---- ALTERNATIVE ROUTE CUTTING OUT PRE-HEAT ROLL
······· ALTERNATIVE ROUTE INCREASING PRE-HEAT TIME

Figure 9.13. Schematic diagrams of complete coating lines

A diagrammatic illustration of a complete coating line is shown
in *Figure 9.13*. Speeds of up to 2.5 m/s (500 ft/min) are obtained
on such units, but speeds in excess of 10 m/s (2000 ft/min) have
been reported[32].

9.5.4 LAMINATING

In addition to normal single-layer coating, the equipment
described above can, with suitable additions, be used for the production of sandwich laminates such as paper/polyethylene/
paper, or even paper/polyethylene/fabric net/paper. Such
laminates require additional unwind stations and preheaters, and
the extra substrates may enter the nip on either side of the plastic
film according to the type of laminate required.

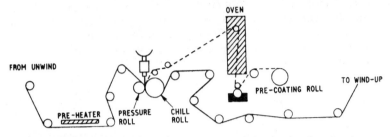

Figure 9.14. Schematic diagram of a Waldron-Hartig laminating line showing second substrate preheating unit

An illustration of the layout of one such laminating line is given in *Figure 9.14*.

9.5.5   SPECIAL EQUIPMENT

As with most other processes, there are many pieces of additional equipment which may be added to a coating line to simplify the operation or improve the performance. Antistatic treatment, thickness measurement and embossing units are only a few of these. Extruding vertically downwards into the nip as opposed to turning the plastic flow through 90° from a horizontally mounted extruder has been adopted by at least one firm and more details have been given in Chapter 5.

9.6   Tube and pipe

The essential units in one form of take-off system for tube are shown in *Figure 9.15* and consist essentially of a cooling trough, a pulling device, and a reeling or cut-off system.

A description was given in Chapter 8 of the way in which the diameter of the tube which issues from the die is controlled by the sizing former or calibrator attached to the front end of the tube die. When the sized tube leaves the calibrator it enters a water bath and is completely submerged during its passage through this bath. The tube enters and leaves the water bath through rubber glands which control the water leakage, although some systems use a further tubular sizing former at the water bath entry in place of the first gland. It is sometimes necessary to divide the water bath into sections by the introduction of intermediate rubber glands so that cooling may be effected in stages.

Figure 9.15. Pipe extrusion line shown in operation producing 600 mm (24 in) Class D pressure pipe. (Courtesy Cincinatti Milacron G.m.b.H., Vienna)

An alternative system of tube production which is particularly suited to small tube and profiles, and already described in the section dealing with dies for such products (Chapter 8), uses an assembly of sizing plates located in the cooling trough and relies for satisfactory working on the draw-down of the product from an oversize die. The dimension of the holes in the sizing plates are made to suit the degree of draw-down as previously described. Diagrammatic illustration of this and of the foregoing system are given in *Figures 9.16a, b.*

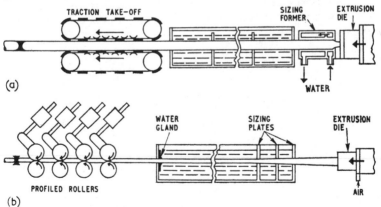

Figure 9.16a, b. Schematic arrangements showing two methods of tube production

On leaving the cooling tank the tube enters a driven nip roll or belt traction device, which pulls it smoothly from the die and through the cooling system. The take-off must be adjustable over a wide, infinitely variable speed range and it must be possible to obtain speed changes smoothly, and without disturbing the running of the equipment. On the efficient functioning of the pulling equipment depends the success of the whole process since any irregularities in the speed of the device are immediately reproduced as ripples in the tube bore. It must also be possible to vary the grip on the tube so that thin-walled sections are not permanently distorted whilst on the other hand ample tension is available to pull heavy-walled pipe through the sizing system. It must be possible to effect this adjustment easily whilst the process is in operation so that the optimum clamping force may be found by trial and error. The pulling device has been the subject of a number of ingenious ideas and many different methods of gripping and simultaneously pulling the tube have been devised.

One of the earliest forms of continuous tube take-offs consisted of a number of pairs of grooved nip rolls in series of which the lower members of each pair were driven whilst the upper rolls were pressure-loaded to grip the pipe as shown in *Figure 9.16b*. This method of take-off has, however, been largely supplanted by later and more efficient systems of the caterpillar type (*Figure 9.16a*).

Another system is presented by the equipment manufactured in Germany by A. Reifenhauser, and consists of three or more endless belts driven at the same peripheral speed and arranged radially in respect to the pipe. The belts can be adjusted to apply pressure to the pipe and are supported on their inner sides, thus ensuring positive drive and a high tractive force. A diagrammatic illustration of this equipment is given in *Figure 9.17*.

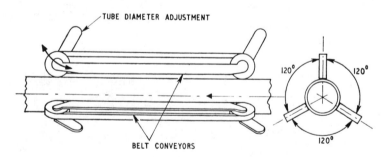

Figure 9.17. Radial belt traction unit. (Courtesy A. Reifenhauser K.G., Troisdorf)

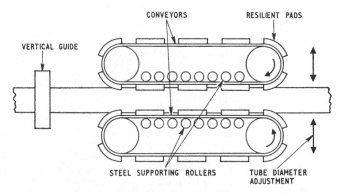

Figure 9.18. Resilient pad caterpillar traction

Figure 9.19. Caterpillar take-off operating on an irregular profile. (Courtesy Mil Ltd., Boston Machinery Division, Wolverhampton)

The caterpillar system, which is now probably the most widely used, as illustrated in *Figures 9.16a* and *9.18,* and contrary to the two methods described above, which are suitable only for pipe, can be used to take-off sections of any form. *Figure 9.19* is a photograph of one such take-off equipment operating on a non-circular section. The take-off can be seen to consist of two endless belts which can be pressurised against the extruded product. On the

Figure 9.20.
A multi-belt traction unit followed by a
planetary cut-off saw and a discharge
table. (Courtesy Speedex (Engineering) Ltd,
Bradford)

Figure 9.21. Automatic extrusion lines for PVC pipe showing cut-off and stacking system. (Courtesy Establissements Andouart, Bezon, Paris)

endless belts are a series of resilient pads which readily conform to the shape of the section. The belts are supported on the contact run by a series of small steel rollers which ensure contact with the extrusion over the full distance between the belt pulley centres. The distance between the belts can be adjusted without disturbing the centre height and a separate adjustment is provided to vary the overall height to suit different extruders.

For the take-off of capillary and small tubing and profiles generally, which usually cannot be handled by the foregoing systems, it is normal practice to use a plain belt conveyor with or without a free-running weighted pressure roller at the far end.

If the tube is flexible, as in the case of semi-rigid PVC or polyethylene, it can be reeled in the same way as described for cable. The reeling may be driven through the central spindle by a slipping clutch device and a variable-speed motor. Alternatively, some form of friction driving band can be adopted which operates on the outer layer of reeled pipe and thus ensures a constant peripheral speed of take-up. The pipe take-off is ideally capable of taking two reels which are mounted side by side so that a fresh reel may be started as soon as the preceding reel is full. It is usual to replug or reseal the tube before it is cut so that the internal air pressure is not disturbed but, in any event, this operation can usually be carried out sufficiently quickly to have little or no effect on the finish of the molten portion.

For large diameter rigid tubes of unplasticised PVC polyethylene and similar materials a cut-off mechanism is necessary since it is impossible to reel these products. A form of circular saw is often used, which can be clamped to the tube and travel with it as the cut-off is carried out. For larger sizes of pipe, planetary saw systems are available. Rotating knives are also used in the same manner. The tube is then replugged in the normal way and the sawnoff length of pipe removed. Automatic pipe and profile handling equipment is available which can be set to cut off the product in accurate lengths and then to stack it in an adjacent 'catcher'. Such equipment is shown in *Figure 9.20*, whilst three lines of pipe equipment of another make is shown in operation in *Figure 9.21*.

Certain complications arise on the tube take-off of large-diameter tubes or pipes and special take-off systems have been developed to cater for these large sections. The difficulties are associated with sealing the pipe in order to maintain the internal air pressures and with obtaining a sufficiently good grip on the pipe to pull it from the die through the cooling system without distortion or surface damage. The diagram in *Figure 9.22* and the photograph *Figure*

8.34 in Chapter 8 show a twin-trolley-type haul-off and illustrate the general principle of this arrangement, which was developed by Tenaplas Ltd, in 1947 and used for large-diameter pipes.

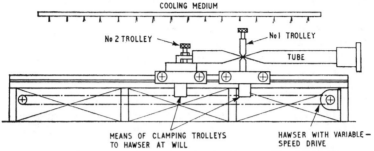

Figure 9.22. Trolley system pipe take-off. (Courtesy Tenaplas Ltd)

Although now primitive, this system is of some historic interest since it illustrates the method by which large-diameter thermoplastics pipes were first produced and shows an ingenious endeavour to convert a manual operation directly into a mechanical system. Moreover, it successfully produced many kilometres of pipe before the system was replaced by later and better methods.

The base of the take-off, which was frequently as long as 18 m (60 ft), formed a double rail or track on which ran two trolleys. Each trolley was fitted with a clamping device and, furthermore, each could be manually coupled or uncoupled, at the correct stage in the operation, to a continuously running wire hawser which passed over a pulley at either end of the take-off, and was driven by a variable-speed motor via a capstan.

The trolley take-off was operated in the following manner.

The soft pipe, when it first emerged from the cooling die, was squashed flat by the large clamps on trolley No. 1, and the molten internal walls adhered to form a complete air seal. Pressurised air could thus be admitted to the pipe through the die in the normal way and was unable to escape. At the same time, by fixing trolley No. 1 to the moving wire hawser, the pipe was hauled along at the requisite speed, supported at regular intervals by idler rollers. Water cooling sprays were located above the take-off system, and the pipe was cooled as it passed under these sprays instead of passing through a water tank as in the previously described system. When it was judged that the crushed pipe had set, trolley No. 2 was attached to the end of the pipe by its clamp and was also coupled to the driving cable. Trolley No. 1 was unclamped from the pipe and the cable and was pushed back along the track towards the extruder whilst the pipe continued to be hauled by trolley No. 2.

When a sufficient length of pipe had emerged from the extruder, trolley No. 1 was again clamped over the pipe near to the die, to form another seal, and was again attached to the driving hawser. As soon as the pipe had cooled sufficiently it was cut off near to trolley No. 1, but on the side remote from the extruder, and when the first nip had been trimmed from the pipe and removed from trolley No. 2, the whole process was repeated.

There was usually a long trough running underneath the first part of the take-off to catch the cooling water flowing off the pipe and frequently a circulation pump was fitted so that the same water could be utilised continuously. In some cases the trolley drive was effected by means of a continuous lead screw instead of the hawser, and it was claimed that ripples in the bore of the pipe due to spring in the hawser were thus eliminated.

The foregoing method for the handling of large pipe, although ingenious and unique at the time of its first use, was wasteful in material as the clamped portions at the end of each tube run needed to be rejected and ground for re-use. This effect obviously reduced with increase in length of the trolley track as two or more standard pipe lengths could be obtained from one tube run with one waste piece only. It could never reduce to zero, however, and the cost of a long track was high both in investment and space requirements. For these and other reasons, this method has now largely been replaced by traction belt take-off systems of the types previously described, and by the vacuum or the trailing plug methods of air pressure retention as discussed in Chapter 8.

9.6.1 DIMENSIONAL CONTROL OF PIPE

The outside diameter of a pipe can be easily measured during extrusion and the necessary corrections applied manually by the operator or the sizing system can, if necessary, be exchanged for one of a different dimension. The wall thickness of the tube and therefore its internal diameter and concentricity cannot be so measured and until comparatively recently the only method available was to cut off a length of the tube and measure it manually. Subsequent dimensional control by a series of such measurements and trail and error adjustment was very wasteful of both time and material. Moreover, there was no guarantee that the dimensions even when so adjusted would not change due to some unnoticed disturbance to produce a further quantity of reject product.

Accurate control of pipe wall thickness is also desirable for another important reason. Limits for the wall thickness of pipes are usually laid down in standards specifications, or are specified by the customer. In order to ensure that the pipe wall is of adequate thickness at all points, it is common practice to run the product on the top limit or even a little over, and this represents a wastage of material which in these days of low profit margins can easily convert a possible profit into a certain loss.

To overcome this problem, systems are now available for the continuous measurement of the wall thickness of non-metallic tubes and pipes by the detection and continuous recording of capacitance changes. If desired, this equipment can also be arranged to control automatically the speed of the take-off to keep the wall thickness constant, and dimensional control to better than 1 per cent without attention from the operator is said to be possible.

By the use of such equipment it is possible to detect roughness of the bore and ripples, etc. whilst the tube is running and to observe the effects of temperature adjustments, etc.

9.6.2 BIAXIALLY ORIENTATED PIPE

An interesting development due to Farbwerke Hoechst in conjunction with A. Reifenhauser, of Germany, is for the production of biaxially orientated high density polyethylene pipe. It is known that polyethylene film and sheet when stretched in two directions under controlled temperature conditions shows greatly increased tensile strength[33]. This led to the development of pipe orientation equipment in order to duplicate these improved strength properties, and thus to save material and costs.

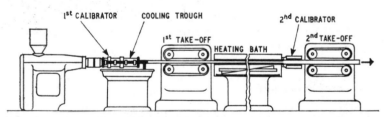

Figure 9.23. Diagrammatic layout of orientation equipment for polyethylene pipe

One method of applying this development is to extrude a thick-walled tube using an external calibrator sizing system and internal air pressure. The cooled tube is hauled off by a caterpillar unit and led into a special heating bath where its temperature is raised by circulated ethylene glycol to approximately 125 °C. The temperature of the bath and its length vary according to the tube dimensions and rate of output. A second calibrator of larger bore is located immediately after the heating bath and the tube expands to fill this calibrator due to the internal air pressure, and is thus transversely orientated. A second caterpillar take-off unit which is adjusted to operate at a higher linear speed compared with the first unit draws the tube through the second calibrator and simultaneously stretches it longitudinally. A diagrammatic layout of the above equipment is shown in *Figure 9.23*.

9.6.3 PIPE PRODUCTION ON SITE

With the increasing use of plastics piping for the conveyance of water, sewage etc., it is becoming apparent that the transportation of finished pipe to remote sites is an expensive and often difficult operation. With this in mind the plastics machinery manufacturing company R.H. Windsor Ltd produced and publicised early in the 1950s a mobile, complete extrusion plant in a truck trailer which could be transported by air for the on-site production of pipes in various thermoplastics materials as required by the particular application[46].

This development, although ingenious, was much before its time and little has been heard of it since. It is interesting to note, however, that a mobile extrusion unit for pipe has been in operation in Scandinavia where it has been used for the laying of pipes in continuous lengths over difficult terrain, and in one case provided a continuous pipe of some 500 m in length which was towed by boat across a small lake and subsequently sunk to the lake bed.

The unit in operation is said to be capable of an output of pipe up to 250 kg/h and the extruder is installed in a standard container for transportation by truck with the generator and power unit mounted in a separate trailer.

9.7 Profile take-off

The take-off methods used for profile work are essentially the same as those for small tube. The section leaves the die of the extruder and enters the cooling trough through one or more sizing plates which support the section at points where it tends to collapse. The sizing plates also form a seal at the end of the trough to limit the water leakage to a reasonable amount. It is not possible to give more than a general description of sizing techniques since each section has to be treated on its merits, and the process is often the subject of patents[34,35]. On leaving the cooling bath, the section passes into a pulling device as previously described, or on to a conveyor band. The rolls of the pulling device may be specially shaped to nip on certain surfaces of the section or resilient pads may be used on a caterpillar unit, but if this is not possible the section is run onto a conveyor band and the haul-off tension is obtained by the frictional force between the section and the conveyor band, which may or may not be assisted by a pressure roll.

Similar reel-up techniques can be applied to sections as can to tubes; thus flexible sections can be reeled, whilst rigid sections are cut off into lengths and stocked in this form.

General note on pipe and profile processes

The above description of processes has been devoted largely to the extrusion of sections in rigid or semi-rigid materials which set-up or become hard on chilling. These thermoplastics, which include the majority, such as acrylics, cellulosics, the polyamides, polyethylene, polystyrene and rigid PVC, lend themselves to sizing by means of water-cooled sizing formers.

Flexible or plasticised PVC is rubbery when cold and does not lend itself to such treatment. The process for the extrusion of tubes and sections in this material, therefore, does not include sizing devices. The extrusion is usually drawn from an oversize extrusion die by means of a variable-speed conveyor belt and a small degree of tension along is usually sufficient to preserve the form. Cooling may be carried out by means of air blast, water sprays, or by a water bath, and final dimensions are controlled by adjusting the tension and thus the amount of draw from the oversize die.

9.8   Monofilament plant

As explained in Section 2.4, dealing with spinning or spinneret extrusion, many filamentous materials are extruded from solution or from a low viscosity melt by means of pumps of substantially normal design in conjunction with a die or spinneret containing a large number of fine holes. Such filaments are generally very fine and are bunched or spun together to give a multifilament continuous yarn which is then 'drawn' or orientated to give the required strength properties. After drawing the yarn may be cut into staple fibre to be spun into staple threads on conventional yarn-spinning equipment.

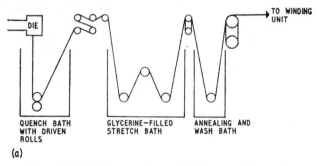

Figure 9.24. (a) Diagram of polyethylene monofilament line

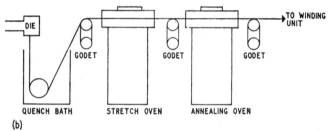

(b) diagram of nylon and polypropylene monofilament line

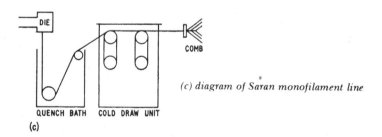

(c) diagram of Saran monofilament line

So far as the plastics industry is concerned, however, and bearing in mind the slit film systems discussed in an earlier section of this chapter, the term 'monofilaments' refers to single thread-like filaments of diameter greater than about 0.09mm (0.0035in) which are handled individually in the manufacturing process and not bunched or spun as in the textile industry. Apart from this difference, the process of producing monofilaments is very similar to that for multifilament textile yarn in that a thermoplastic is extruded downwards through a multi-orifice die plate into a cooling system or quench bath. After cooling, the filaments are also drawn—but in this case separately—at controlled temperatures when they become greatly reduced in section and acquire increased tensile strength.

Polyethylene, polystyrene, nylon, Saran, PVC and polypropylene are all extruded into monofilaments and diagrammatic illustrations of three types of extrusion line for different materials are given in *Figure 9.24a-c*. Monofilament sizes range from about 0.09 up to 1.5 mm (0.0035 up to 0.060 in) and larger in diameter, and are generally of circular cross section, but other profiles have been developed.

It is very difficult to give a general picture of the type of equipment required for monofilament production as much of the development work has been concerned with specific requirements, and so much depends on the material and its properties. Different materials require different treatment. Saran, for example, requires no further heating after the quench bath, whereas nylon and polyolefins do require this in different degrees. Polystyrene monofilaments, on the other hand, are preferably stretched on heated rolls. Jack and Horsley[36] have described a process for Saran monofilament production and state that a 4 : 1 stretch ratio in the quench bath followed by a further 4 : 1 cold stretch is employed. Groel and Versteeg[37] have given a detailed description of the production of polyethylene monofilaments and the use of ovens instead of heated baths is advocated by Mahn and Vermillion[38] in their description of polypropylene monofilament production.

In most cases the monofilaments are extruded downwards into a quench bath from a die containing up to about 40 holes whose individual dimensions may be up to five times that of the finished filament diameter. The temperature of the bath is controlled according to the requirements of the material and the holes in the die are countersunk or streamlined on their entry to leave a land length of about 3 : 1. In the quench bath the monofilaments pass beneath a free-running roll which can be raised and lowered to facilitate threading, and also to control the filament quench time. A nip roll system is in some cases substituted for the free-running hold-down roll.

After leaving the quench bath the filaments are passed around a pulling roll system, usually consisting of a pair of rolls, which is known as a 'godet' or 'snubber', before entering the stretching or orientating heater. This can consist of either a heated bath, an oven or heated rolls and the method used in any particular case is

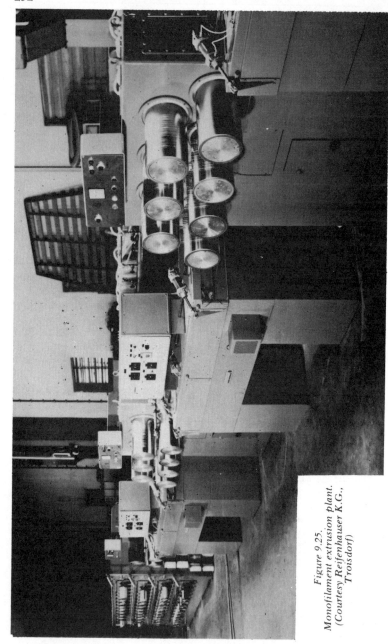

Figure 9.25.
Monofilament extrusion plant.
(Courtesy Reifenhauser K.G.,
Troisdorf)

determined by the required orientating temperature and the properties of the material; in general the temperature must be controlled within close limits.

A second godet or some other form of capstan located immediately after the orientating heater is adjusted to run at a controlled higher speed than the first godet, and this speed differential provides the necessary orientation by stretching the monofilaments whilst they are in the heater. The speed ratio between the two pulling systems is variable from about 2 : 1 up to 10 : 1 and higher, and the degree of stretching is again determined by the properties of the material. The stretching unit including the godet can be repeated if necessary or desirable.

From the stretching unit the monofilaments are drawn through an annealing system, and if necessary a wash bath by means of a further godet running at the same peripheral speed as the stretch godet, and thence to a wind-up unit. The annealing system is usually a controlled-temperature liquid bath, but ovens are also used.

The wind-up unit may take one of several forms, depending on the properties of the material, the thickness of the monofilaments, and their end use. A large drum known sometimes as a 'scrap' roll may be used to wind all the filaments together for sale in cut lengths, or in the case of thicker bristle fibres which would take a permanent set if coiled on a circular drum, a special winder consisting of four posts may be used. Monofilaments which are required in continuous lengths are wound individually on separate spools, each with its own drive and tension-sensitive speed control.

Figure 9.25 is a photograph of a complete monofilament plant in which most of the features discussed above can be clearly seen.

9.9 Coextrusion processes for sheet and film

9.9.1 GENERAL

Coextrusion is the name now given to the comparatively ancient process of combining several different materials into one composite product whereby the advantageous properties of each separate material are combined. Composite sheet and film products are frequently produced in this way which is in some respects—but not all—superior to the adhesive lamination of pre-extruded web products, which is the other important method of producing such composites.

Certain basic information on coextrusion techniques was given in the previous chapter dealing with the design of extrusion dies.

It only remains, therefore, to provide brief details on some of the processes and principles involved.

9.9.1.1   Laminating position

In general there are two basic methods for the production of coextruded film and sheet materials depending on the point in the system where the individual materials are brought together to be combined. This point can be either within the die body just prior to the die land, or after emergence and, therefore, exterior thereto. The former method gives greater scope for the combining of a large number, i.e. 3, 4, 5 and more materials and is the more straightforward in concept, but relies largely on mutual compatibility and material affinity of the materials to give a laminate bond. The latter method gives scope for improving the bond strength by the use of adhesion-promoting agents between the layers immediately before they are brought together, but is more complicated in that several separate die orifices are necessary and is usually restricted to two or at the most three layers.

Of the above two combining systems the former method is by far the most widely used but there is at least one important process which uses the latter system, as will be referred to later.

9.9.1.2   Coextruded sheet

Multilayer sheet products which may contain up to five or more layers involving five or more extruders attached to the one die system are normally produced horizontally using substantially standard sheet take-off units as previously described, and normally—but not always—using the die interior combining point (*Figure 9.26*).

9.9.1.3   Coextruded film

Multilayer film may be produced either flat using one or other of the flat film processes previously described, or by the use of a tubular film system. Both the die interior and the exterior systems are in use although the former is the more popular for reasons previously given.

An interesting tubular process for the production of multilayer film uses a multi-orifice tubular film die and relies on the divergence of the bubble which is promoted by the expanding effect of

Figure 9.26. Photograph showing a complete four-layer sheet coextrusion plant. (Courtesy Reifenhauser K.G., Troisdorf)

the internal air pressure, to combine the film materials whilst in the molten state at a point just above the die land exit. Ozone or other gaseous oxidising agents are passed continuously through the die between the orifices thereof to promote adhesion between the layers[47].

9.9.1.4   Coextrusion coating

The extrusion coating system described in a previous section of this chapter can obviously be extended to include coating with multilayer coextruded materials. Apart from the complication in operation and the space requirements due to the need for two or more extruders, few difficulties are encountered and the composite materials so produced can be adapted, depending on the polymers used, to give superior barrier or other properties as required.

9.10   Cellular extruded products

9.10.1   GENERAL

Mention has been made in a previous chapter of processes for the production of expanded or foamed extruded products by the incorporation of suitable blowing agents into the material either before or during extrusion.

The majority of thermoplastics materials can be extruded in a foamed or expanded form and into a very wide variety of products by one or other of the above methods, and the products so produced have the advantages of lightness, strength and low usage of plastics raw materials combined with greater bulk, compared with unfoamed products, and reasonable surface finish.

In general the extrusion processes involved in the production of foamed plastic extruded products require the use of a cooled calibration system to contain the expansion of the product and to provide a good surface finish and an entire 'skin'. Thus the process is to some extent analogous to the non-expanded pipe and profile extrusion systems previously described in which internal air pressure rather than internally evolved gas provide sliding pressure contact with the calibration system[48]. Other methods have been developed and described[49].

Figure 9.27. Photograph showing expanded polystyrene extrusion equipment in operation using a 150 mm (6 in) extruder and slitting system.
(Courtesy N.R.M. Corpn, Akron, Ohio)

9.10.2 EXPANDED POLYSTYRENE

A possible exception to the above thoughts on the extrusion of expanded products arises in considering processes for the manufacture of expanded polystyrene sheet and similar products using liquid hydrocarbon or Freon blowing agents which are incorporated into crystal polystyrene granules either before processing or during extrusion.

In one such process pentane is used as the blowing agent and melt temperatures of the order of 115-130 °C are required. Relatively deeply cut short-compression, short-metering-zone screws with 2.5 : 1 or 3 : 1 compression ratios are normally used in order to avoid excessive overheating due to shear, and screw speeds are kept low.

In operation the material is brought to the required melt temperature in the compression zone of the screw and then cooled in the metering zone and die. It is an important feature of extruders used in this work therefore that efficient cooling means be provided for the metering zone.

The product is normally produced as a thin-walled tube on a horizontal form of tubular film system, as previously described. After it leaves the die the tube diameter increases until the expansion is completed, after which the product may be collapsed and flattened at the take-off and edge-trimmed into two-sheet thicknesses, or slit and opened (*Figure 9.27*).

Several other systems exist for the manufacture of expanded polystyrene sheet and these include the use of twin-screw extruders; two single-screw machines in series, the first being for mixing and gas injection, the second for cooling; single-screw machines with blowing agents injected into the barrel vent by diaphragm pumps, and others. In most case, however, the take-off systems used are as previously described.

Similar systems have been employed in the production of foamed sheet in the polyolefin materials and controlled densities between 25 and 100kg/m³ (1½ and 6 lb/ft³) have been successfully produced.

REFERENCES

1   PATON, J.B. et al., 'Extrusion', in *Processing of Thermoplastic Materials*, Ed. Bernhardt, E.C., Reinhold, New York (1959), p. 288
2   RHODES, P.H. and BLACK, H.C., 'Continuous Extrusion and Thermal Curing of Insulation', *S.P.E. tech. Pap.*, **5** (1959)
3   ZIMMERMAN, W.F., 'Insulated Wire Manufacturing Costs Trimmed with Diameter Control Gauge', *Wire & Wire Prod.*, **35**, 601 (1960)

4   HARRINGTON, A.W. and KELLEY, H.L., 'Synchronised Extruder—Capstan Drives for High Speeds', *Wire & Wire Prod.*, **35**, 1658 (1960)
5   ENTWHISTLE, J.L., 'Modern Developments in the Control of High Speed, Low Tension Reeling Equipment', *Wire & Wire Prod.*, ·**35**, 1664 (1960)
6   LYNCH, A.G. and SPEIGHT, E.A., 'A Photoelectric Device for Measuring Cable Diameters During Extrusion', *Proc. Instn elect. Engrs*, 109, Part A, Sup. No. 3 (1962)
7   HAINE, W.A. and LAND, W.H., 'Effect of Extruder Variables on Properties and Output of Polyethylene Film', *Mod. Plast.*, **29**, 109 (1952)
8   HAINE, W.A. and ROBINSON, H.B., 'Extrusion of Polyethylene Film with Controlled Properties', *S.P.E. tech. Pap.*, 4, 738 (1958)
9   DOYLE, R., 'Extrusion of Clear Film from High Density Polyethylene', *S.P.E. J1*, **14**, 35 (1958); DOYLE, R. and DETTER, C.V., 'High Speed Production of Clear Film from High Density Polyolefins', *S.P.E. J1*, **15**, 1079 (1959)
10 MEYER, F.J., 'How to Make Roll Chilled Polyolefin Films Commercially', *Mod. Plast.*, **36**, 97 (1959)
11 JACK, J., 'Biaxial Stretching of Polypropylene Film', *Br. Plast.*, **34**, 312, 391 (1961)
12 COLLINS, F.H., 'Controlled Density Polystyrene Foam Extrusion', *S.P.E. tech. Pap.*, **6** (1960)
13 PLAX CORPN, U.S.A., U.S. Pat. 2 559 386 (27 October 1948)
14 VISKING CORPN, U.S.A., U.S. Pat. 2 461 975 (20 October 1945)
15 VISKING CORPN, U.S.A., U.S. Pat. 2 461 976 (20 October 1945)
16 VISKING CORPN, U.S.A., Br. Pat. 716 160 (29 September 1952)
17 VISKING CORPN, U.S.A., U.S. Pat. 2 542 652 (6 December 1946)
18 BRITISH XYLONITE CO. LTD., Br. Pat. 710 271 (22 May 1951)
19 CLEGG, P.L. and HUCK, N.D., 'Effects of Extrusion Variables on the Fundamental Properties of Tubular Polyethylene Film', *Plastics, Lond.*, **26**, 114 (1961)
20 GRANT, D., 'Current Advances in Film Manufacture', *Plast. Prog.*, 151 (1961)
21 PILARO, J.F., KREMER, R.J. and KUHLMANN, L.A., 'What's the Best Blow Up Ratio?'. *Mod. Plast.*, **39**, 123 (1961)
22 ZUKOR, L.J., 'The Annealing Chamber for Blown PE Film', *Plast. Technol.*, **5**, 52 (1959); *Br. Plast.*, **33**, 11 (1960)
23 VAN DER HEIDE, J.C. and WILSON, J.L., 'Guide to Corona Film Treatment', *Mod. Plast.*, **38**, 199 (1961)
24 KREFT, L. and DOBOCZKY, Z., 'Oberslachenvorbehandlung von Kunststoffhalbzeugen', *Plastverarbeiter*, **13**, 169 (1962)
25 E.I. DUPONT DE NEMOURS & CO. INC., U.S.A., Br. Pat. 738 474 (18 November 1953)
26 BRITISH CELLOPHANE LTD, Br. Pat. 723 631 (7 July 1953)
27 KREIDL, W.H., U.S. Pat. 2 632 921 (18 January 1949)
28 KRITCHEVER, W.F., U.S. Pat. 2 648 097 (4 April 1952)
29 BLOYER, S.F., 'Treating Polyethylene for Printing', *Mod. Plast.*, **32**, 105 (1955)
30 VISKING CORPN, U.S.A., Br. Pat. 722 875 (27 January 1952)
31 VISKING CORPN, U.S.A., Br. Pat. 715 914 (27 February 1952)
32 E.I. DEPONT DE NEMOURS & CO. INC., U.S.A., 'Du Pont Conducting Resin and Mechanical Research Aimed at Big Increase in Extrusion Coating Speed', Du Pont Product Information Service
33 RICHARD, K., DIEDRICH, G. and GAUBE, E., 'Strengthened Pipes from Ziegler Polythene', *Kunststoffe*, **50**,(7), 371 (1960)

34 SLAUGHTER, C.E., U.S. Pat. 2 663 904 (12 December 1950); Br. Pat. 603 077 (9 June 1948)
35 E.I. DUPONT DE NEMOURS & CO. INC., U.S.A., Br. Pat. 705 789 (16 April 1952)
36 JACK, J. and HORSLEY, R.A., 'The Extrusion and Properties of Saran Monofilament', *J. Appl. Chem.*, **4**, 178 (1954)
37 GROEL, J.F. and VERSTEEG, J.H., 'Polyethylene Monofilaments', *S.P.E. tech. Pap.*, **4**, 523 (1958)
38 MAHN, H.M. and VERMILLION, J.L., 'Polypropylene Monofilament Extrusion', *Plast. Technol.*, **7**, 23 (1961)
39 S.A.S. LAVORAZIONE MATERIE PLASTICHE L.M.P. DI M.I. COLOMBO & C., Br. Pat. 825 392 (28 March 1958)
40 BONNER, R.M. 'A New Take-Off Method for Manufacturing Acrylic Sheet', *S.P.E. Jl.*, **19**, 1157 (1963)
41 ANON., 'Iceator Blown Film Cooling System', *Br. Plast.*, **43**, 131 (1970)
42 ANON., 'Mandrel Cooler For Higher Film Outputs', *Europlastics Monthly*, 101 (June 1972)
43 Schyeldahl, East Providence, U.S.A., Maker's literature
44 GENERAL ENGINEERING CO. (RADCLIFFE) LTD., Br. Pat. 1 119 552 (9 June 1966)
45 ANON., 'For better polypropylene film; cool it', *Mod. Plast.*, **47**, 104 (1970)
46 ANON., 'Pipe made at point of use', *Mod. Plast.*, **32**, 117 (1954)
47 ALKOR-WERKE KARL LISSMANN KG, Munich, Br. Pat. 920 159 (5 July 1961)
48 STE. UGINE KUHLMANN, Fr. Pat. 1 498 620
49 NIGHTINGALE, R.J., 'Cellular profiles: the wood of the future', *Europlastics Monthly*, 41 (February/March 1974)
50 I.G. FARBEN INDUSTRIE, Br. Pat. 479 202 (2 February 1938)
51 RASMUSSEN, O.B., U.S. Pat. 2 954 587, (23 May 1955) 2 948 927 (29 April 1957)
52 THE SHIRLEY INSTITUTE, Didsbury, Manchester, Report issued in 1964 discussing the 'Polystress' process
53 SMITH & NEPHEW RESEARCH LTD., Br. Pat. 914 489 (14 July 1961)

10

Other Important
Plastics Processing Techniques

10.1 Introduction

Many well known thermoplastics processes rely on an extrusion system to provide the heat-softened material for manipulation into a final finished article. Injection moulding is an obvious and possibly the oldest example of this where a thermoplastics material is forced by a ram through a heating chamber and thence to a nozzle or die and finally into a closed mould where the material takes up the required form. Transfer moulding is another example in which a plastics material is extruded into a mould.

The foregoing, however, are examples in which a ram extruder is used as the operative member. With the greater understanding of screw extrusion mechanisms which has accrued, and the obvious advantages of better heating and mixing which such systems offer, the ram system has been largely replaced by screw mechanisms in all but the very smallest sizes of machine.

Moreover the use of a screw extruder to provide a continuous supply of homogeneous melt has enabled a number of processes which previously were discontinuous, and often slow, to operate in a substantially continuous manner.

10.2 Injection moulding

Injection moulding consists essentially of melting a thermoplastics material and then forcing it under high pressure into a closed mould. In the original ram type machine the melting occurred whilst the material was being forced by a ram or piston through a number of fine channels in an externally heated 'heating chamber' on its way to the nozzle and thence to the mould (*Figure 10.1*).

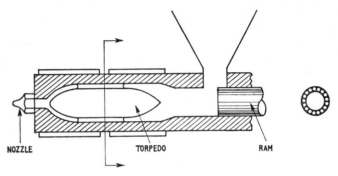

Figure 10.1. Diagrammatic illustration of a ram-type heating chamber

Unfortunately the conditions necessary for uniform heating and for high pressure at the nozzle are contradictory in such a system and one requirement must always be sacrificed for the other. On the one hand in order that the heating chamber should effectively plasticise the material the internal channels should be small in cross section and as long as possible. On the other hand, in order to minimise pressure drop in these channels they should be as large as possible and of short length.

10.2.1 TWO-STAGE PROCESSES

One way of overcoming this problem is known as the two-stage system in which the material is melted in a separate chamber and then passed to a pressure chamber for injection. The melting chamber or preplasticiser as it is called can either be a heating chamber as described above or a screw system may be used. An important advantage of screw systems is that owing to the method of heating employed, i.e. internal friction and shear, a melt suitable for injection can often be obtained at lower temperature, avoiding high temperature gradients and thus simplifying the moulding of heat-sensitive materials such as unplasticised PVC.

The use of screw systems to provide a melt either for direct extrusion into a mould or to supply a second chamber for injection by a ram at high pressures is now old history. The first patent[1] on such a process was granted in Germany to Eckert and Ziegler in March 1927, and used a screw system—twin screws in this case—for direct extrusion into a mould without any form of ram pressurising.

Figure 10.2. British machine for direct extrusion into a series of moulds on a rotating table. (Courtesy Foster, Yates & Thom Ltd, Blackburn)

The single-screw low pressure rotary table injection machines which were produced in the U.K. by Foster Yates & Thom[2] and others, and in Germany by Desma-Werke, Paul Troester[3] and others, used this principle for the production of vinyl footwear and shoe soles and other thick-section mouldings (*Figures 10.2-10.4*). The use of a screw extruder for the extemporaneous production of a small number of mouldings, by means by a simple split mould clamped to a special nozzle die and extracted by hand, has been practised since 1945 or earlier[4].

The next important patent on a screw system for injection moulding was granted to the Celluloid Corporation of New Jersey in 1935 and consisted essentially of a screw preplasticiser feeding into a ram pressure chamber, thence to the mould[5]. Certain of the larger present-day injection machines employ systems which are substantially similar to this early design. An ingenious extension of the same method employed a hollow screw in which the injection ram operated. Thus the screw and the ram were concentric and in line and several variations of this idea have formed the basis for commercial injection machines[6,7].

10.2.2 SCREW RAM SYSTEMS

Perhaps the most important step in the utilisation of screw systems for injection moulding occurred in 1943 when Hans Beck of

Figure 10.3. A modern version produced in Germany of the type of machine shown in Figure 10.1. (Courtesy Desma-Werke G.m.b.H., Achim bei Bremen)

Figure 10.4. German machine for direct extrusion into a multi-mould system. (Courtesy Paul Troester, Hannover-Wulfel)

Badische Anilin und Soda Fabrik filed a patent covering several such systems, one of which used a screw which could be moved axially to act as a ram[8]. In operation, the rotating screw, which commences in the fully forward position, takes material from the hopper, melts it and forwards it towards the nozzle of the machine, which at this stage in the cycle is closed by some form of valve arrangement. The melt, which is forced into the space between the closed nozzle and the screw termination, causes the screw to move back along the barrel, usually against a light hydraulic pressure, until at a certain predetermined position an electrical contact is made, stopping the screw rotation. At the same time the hydraulic ram system is initiated, forcing the screw forward to inject the material into the mould and the nozzle valve is opened either by pressure against the mould or by means of a separate hydraulic cylinder and suitable linkage.

A large proportion of the screw injection machines commonly available at the present time are based on the axial screw system described above or on some modification thereof. The system generally works extremely well and with material of high viscosity such as rigid PVC there seems to be no appreciable loss of injection pressure due to the material being forced back along the screw channel rather than into the mould. With low viscosity melts, however, some more positive pressure system is required and several methods of obtaining this have been devised, and patented. These methods include non-return valves of various types at the end of the screws[9], piston heads to the screws engaging with closely fitting cylinders in the machine heads[10][11] and intermeshing twin screws[12].

There is, however, another problem associated with the use of a screw as an injection means which causes loss of injection pressure and this is 'screw drag'. When a screw whose flights are filled with material is moved axially in a barrel, a considerable resistance is offered to the movement by friction of the material against the barrel bore. Part of the energy available for injecting the material into the mould is, therefore, dissipated in overcoming this friction. The friction obviously increases with pressure and in screws which have no non-return valve or other system of isolating the injection material from that in the screw flights is therefore greatest during the injection stroke. This problem does not arise to the same extent with valved or plunger headed screws and not at all with systems in which separate injection chambers are used.

10.2.3 CONTINUOUS OPERATION

The screw injection processes described so far are discontinuous in operation and are thus unable to take full advantage of the screw extrusion mechanism. Despite the more uniform heating, better mixing and lower-temperature melts which result from the use of screws, the need to stop the preplasticising operation during the injection stroke prevents the full utilisation of these features and sets a limit to the melt homogeneity which can be obtained. The present trend, therefore, in injection machine design is towards increased continuity and at least one system has been developed in which the screw operates continuously[13].

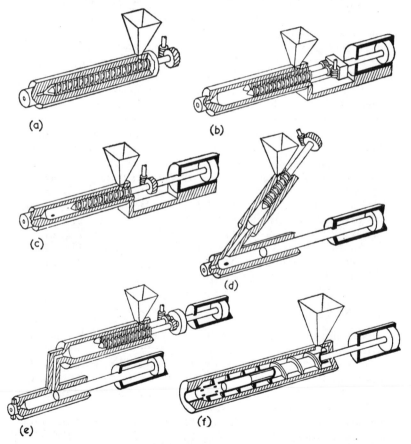

Figure 10.5. Diagram illustrating various types of screw preplasticising

The various ways in which screws can be used in injection moulding processes are illustrated diagrammatically in *Figure 10.5* and discussed at length elsewhere[14].

10.2.4 CONSTRUCTIONAL FEATURES

Screws and barrels for use in injection moulding systems generally follow the principles outlined in other chapters of this work, but due to the rather more severe conditions under which such components operate there are several features which require special attention in design.

The barrel of a normal extruder although frequently designed to withstand a test pressure of 150 MN/m^2 (20000 lbf/in^2) is seldom, if ever, expected to experience such pressures in use. In an injection moulding machine cylinder on the other hand, these and higher pressures are quite normal. This obviously means that the barrel for an axial screw injection system must be designed to withstand these pressures in normal working, which calls for great care in the design of the main barrel and particularly the liner and in the selection of materials for those members. It is important to ensure that the hoop stress of the barrel assembly is within the elastic limits of the liner material used and because of this it is common practice to shrink the liner into the barrel to preload the liner material.

In the case of a screw to be used as a ram in an axial reciprocating system, it is important to ensure that the column strength is adequate to withstand the end loading during injection and that the torsional strength at the feed zone is suited to the torque load during the plasticising cycles when the screw might be required to start and stop from full speed several times per minute. The thrust-bearing system also must be adequate to give reasonable service life under these arduous conditions.

10:3 Blow moulding

Another important plastics fabrication process which relies on extrusion of one form or another to supply a melt in the correct form and condition for final manipulation is blow moulding. By this means thermoplastics materials are formed into hollow articles such as bottles and containers, dolls and other hollow toys. Recently the scope of this technique has been extended to include the production of articles such as buckets and bowls, refrigerator shells and liners, pipe fittings, etc., in direct competition with other processes such as injection moulding.

Basically the blow-moulding process consists of extruding a tube of molten thermoplastics material known as a 'parison' into a mould of required form, inflating the tube to contact and set-up against the cooled walls of the mould cavity and then extracting the finished or semi-finished product.

This simple concept is, however, much complicated by a multiplicity of major or detail modifications, through the variation of one of more features of the process, which has resulted in an extremely complex list of variables and a difficult and formidable patent situation. As this subject has been dealt with at length elsewhere[15,16], this section will be concerned only with the requirements of extruders for the blow-moulding process and with other matters of especial interest.

In general an extruder for the production of parisons for blow moulding must follow quite closely the requirements for high quality extrusion which have been outlined in other sections of this work. The extruder must be quite free from surging or pulsation effects in order to ensure a parison of uniform cross section. It must also deliver an extrudate quite free from surface blemishes and at controllable temperatures. The extruder may also be required to deliver such high quality extrudates intermittently, in which case there must be no variation from one parison to the next or within the length of individual parisons.

10.3.1 MELT TEMPERATURE CONTROL

Another factor which is perhaps more critical in blow-moulding operations than with normal extrusion processes is the temperature of the melt or 'stock'. High temperature low viscosity melts usually result in blow-moulded articles with superior surface finish and strength properties. If the temperature is too high, however, the parison has a pronounced tendency to elongate as it leaves the die producing parisons which are thinner near to the die and thicker at the end remote from it—a phenomenon known as 'necking'. Such parisons are obviously useless for blow moulding because of the pronounced variations in thickness which would occur in the finished product. High melt temperatures also tend to prolong the overall production time because of the longer cooling cycle required and tend to increase the shrinkage of the finished product on cooling. In the case of heat-sensitive materials such as rigid PVC and, in fact, with other materials also, there is obviously greater risk of thermal degradation when the melt temperatures are high.

Low melt temperatures, on the other hand, overcome the problems outlined in the foregoing but if too low will result in inferior products with poor finish and low strength properties although a high blowing pressure is sometimes used to minimise or to disguise these effects. The use of very accurately controlled melt temperatures at or just above the crystallite melt point in crystalline polymers has however been used to give an orientated structure to the material and higher strength[17].

For best results, therefore, as regards appearance, strength properties and rate of production, an extruder for blow moulding must be designed to deliver a melt at a controllable temperature so that the lowest possible temperature commensurate with good quality products can be obtained and held throughout a run and repeated for subsequent runs on similar blown items. The melt, even when comparatively low temperatures are used, must also be delivered at high throughput rates.

These apparently conflicting requirements are best satisfied in practice by the use of a short-metering-zone deeply cut screw of at least 20 : 1 L/D ratio with a short transition zone and rotating at relatively low speeds. The pressure and thus the melt temperature and viscosity would be controlled by means of a suitable valve in the head and for best results it is necessary to include a melt thermocouple in the head system and it is desirable to provide a pressure gauge also.

For heat-sensitive high viscosity materials the screw would be of similar design as regards L/D ratio and depth, but with a continuously tapered transition zone and no metering zone.

10.3.2   HEAT SENSITIVE MATERIALS

A problem associated with the blow moulding of heat-sensitive materials such as clear rigid PVC materials, is concerned with the design of the parison-forming head, which on normal blow moulding equipment for polyolefins, etc., is of crosshead form. The possibility of material hold-up and consequent degradation in dies of this form, particularly under conditions of intermittent operation, is very great so that such dies are often unsuitable for use with such heat-sensitive materials.

Dies with 'bottom' feed in which the side feed limitation is avoided have been used with some success, but even here the need for a 90° elbow also gives problems due to the different flow path lengths on the exterior as against the interior radius of the bend. The best solution available so far, therefore, appears to be in the use of either a horizontal system of moulds or an extruder constructed to operate vertically downwards. Either of these methods would allow the attachment of straight-through parison heads and would thus avoid completely all the problems associated with change of direction of the melt.

10.3.3 LARGE BLOWN MOULDINGS

In the production of large blown articles such as carboys, casks, large hollow toys and similar items requiring a large and heavy parison, the problem of necking becomes very serious despite careful attention to melt temperature and fluidity. Methods have therefore been devised to minimise this difficulty and one of the most important of these is described in an Italian patent by Pirelli[18].

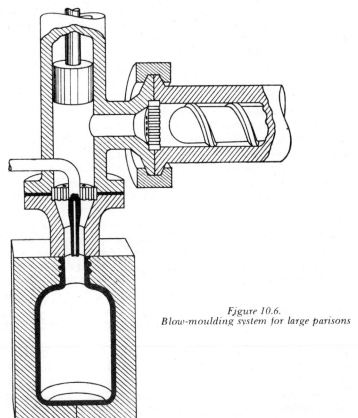

Figure 10.6.
Blow-moulding system for large parisons

The Pirelli system, which is designed specifically to enable heavy parisons to be formed rapidly and thus to reduce 'necking', is basically a two-stage process somewhat similar to a screw preplasticised injection-moulding system. The screw feeds molten materials into the side of a vertical crosshead which has a parison-forming orifice at its lower end and a cylinder with a hydraulically operated ram at its upper end. During operation the molten material from the extruder screw is forced into the crosshead and pushes the ram upwards until the required full charge has accumulated in the cylinder. The ram is then forced down and the parison is formed at a high linear speed. A diagrammatic illustration of such an apparatus can be seen in *Figure 10.6*.

10.3.4 PARISON CONTROL

A further matter which is of considerable interest in the production of blown articles by extrusion techniques relates to the control of parison thickness. If a blown container or other hollow object is required to have a varying wall thickness without a corresponding change of diameter then it is apparent that the parison thickness must also vary. The orthodox extrusion systems from a standard crosshead or parison forming head do not allow such variation but there are at least two methods which have been devised and patented for this purpose.

The first method[19] uses an axial movement of the parison die mandrel to produce a variation in the area of the die orifice and thus to produce corresponding local variations in the parison extrudate thickness. By means of cams and a suitable control system this method can be extended to produce a cyclic regime of thickness variations in each parison as it is formed.

The second method consists of drawing the parison away from the die at a varying rate by a gripping means independent of the mould system whilst simultaneously inflating the parison to contact the mould cavity[15]. Thus the parison can be stretched and thinned locally and again a cyclic sequence can be applied[20].

Finally, on this question of blow moulding, it will be interesting to mention briefly two-colour and two-material containers.

In Chapters 8 and 9 the various methods of obtaining multi-coloured extrusion from one die, or of two or more concentric extrusions of different materials, using two or more extruders attached to the same die system were discussed. These processes are obviously capable of adaption to the production of parisons for blow moulding and containers manufactured by such systems have been produced[21].

10.4 Continuous thermoforming

Vacuum and other systems for the thermoforming of thermoplastics sheet materials have become important for the manufacture of hollow containers such as drinking beakers, ice cream containers, pot closures, frozen food containers, etc. An important feature of such processes is that extremely thin-walled hollow articles which would be impossible to make by injection moulding can be readily formed and because of their low material content and low cost are suitable for disposable packaging.

10.4.1 BASIC PROCESS

In basic essentials the process of thermoforming consists of clamping a sheet of thermoplastics material above a mould, heating the sheet until it becomes softened and formable, and then applying fluid pressure to the sheet to cause it to form closely to the mould shape. There are many variations of this simple concept depending on the shape, size and complication of the article to be formed and the method of applying the fluid pressure.

Compared with processes such as extrusion and injection moulding, the thermoforming processes have one obvious limitation. Expensive sheet materials which already contain the cost of one converting process are used instead of the cheaper raw material in granules. Although this limitation is not of great significance in the production of special-shaped articles in comparatively small numbers—an application to which thermoforming processes are particularly well suited—it does assume great important in the manufacture of disposable containers in very large quantities and at the lowest possible unit cost. This has led to the development of in-line processes in which hot sheet is taken directly from the extruder and formed continuously.

10.4.2 CONTINUOUS PROCESS

Such processes have in general taken two forms. In the first the sheet is taken directly from the extruder onto a conveyor which feeds it to a reciprocating mould carriage containing a large number of mould cavities. The sheet is formed by the mould carriage which then lifts and returns for the next cycle, whilst the sheet passes on to the next station for continuous blanking out of the formed articles. Using a process similar to this, the production of

63 mm (2½ in) square containers, 60 mm (2⅜ in) deep at the rate of 288 per minute was reported[22] in 1958 and more recently the **KLM** Company of Stratford, Connecticut, has developed and successfully commercialised a similar process for the production of screw caps for bottles at the rate of 720 per minute or more depending on the machine's size[23].

The second system uses a wheel rotating in a vertical plane to support the mould cavities which are disposed around its periphery, and thus avoids the relatively complicated mould carriage reciprocation mechanism.

In both systems the waste plastics net which remains after the formed articles have been blanked out is fed directly into a scrap grinder for re-use and may be fed back automatically to the extruder hopper by means of a pneumatic or other conveying system.

10.4.3 CONTINUOUS THERMOFORMING OF BOTTLES

An interesting development from the second of the above continuous thermoforming processes, i.e. by means of a rotating wheel, uses two sheets of thermoplastics material to form closed containers or bottles continuously.

In this process two sheets of thermoplastics material are extruded simultaneously from a multiple die in a vertically downward direction. From the extruder the sheets feed tangentially on to a vertically rotating multi-mould system which is located directly beneath the die. The mould halves in this multi-mould system are arranged to open and close in a horizontal direction and in closing to trap the two hot sheets and to seal the edges. Vacuum is applied immediately to the mould half cavities to form a hollow article, after which the moulds open. This process is repeated as each mould reaches the correct position beneath the die and the containers are continuously removed as a string attached neck to tail[24].

10.5   The preparation of materials for extrusion and injection moulding

In section 5.4 a description was given of the various extrusion systems used for the preparation of a compound for subsequent processing. The presentation of such compounds in suitable granular form for feeding to extruders and injection machines was not dealt with however.

The shape, size and uniformity of the granules often has a considerable influence on the processing properties of thermoplastic materials and this factor should never be overlooked, particularly in the recording of operating conditions on a particular job. A later attempt to produce the same product using the same operating conditions may result in complete failure if the granular form of the feed material is substantially different.

In general there are four systems in common use for the production of granules from thermoplastics materials and each of these may be used in line with the extruder on which the compound has been prepared.

10.5.1 ROTARY CUTTERS

The rotary cutter, which was the earliest method used for the production of plastics granules, consists of a series of knives, attached to the periphery of a massive rotor, which engage with one or more fixed knives arranged to protrude radially inward from the walls of a cutting chamber, within which the rotor revolves (*Figure 10.7*).

Figure 10.7. A large rotary cutter open to show the rotary and fixed knives and part of the screen. (Courtesy U.S. Industries Inc. Engng Co. Ltd)

The base of the cutting chamber consists of a robust screen which is interchangeable to vary the mesh size.

In operation, the plastics material, which may be fed from the extruder as a continuous strip via a water-bath cooling system, is fed into the upper part of the cutting chamber where it is caught between the rotating and fixed knives and cut into random pieces. The pieces then rotate in the chamber and are successively cut and re-cut until such time as the material is sufficiently reduced to pass through the screen openings.

The granules produced by this method are random in shape and vary in size from 'fines' or dust up to the maximum which the screen will pass. The bulk density of such granules is low, therefore, and the heat transfer properties are non-uniform.

Despite the above limitations the rotary cutter still remains one of the most popular methods for the production of granular materials. This is partly due to the fact that these cutters do not need to be fed with uniform strip material, but will also accept scrap materials in any form. As all extrusion and injection operations produce scrap in some proportion the installation of one or more rotary cutters is essential to an economic operation. Moreover, some operators would prefer to have all their material including the scrap in one physical form, rather than cope with two sets of processing conditions for the same product.

10.5.2 CUBE CUTTERS OR DICERS

The second method for the preparation of granules is known as cube cutting and in this process special machines are used to produce cubes of regular dimensions from cooled strip of uniform and suitable thickness directly from an extruder as a continuous operation.

There are various designs of cube cutter but the two forms most widely used are known as 'slitter-type' cube cutters and 'stairstep dicers' respectively.

In the slitter type the strip of material is first slit into 'laces' by means of a series of rotating disc knives. The laces are then fed to a rotating cutter operating against a fixed knife where they are cropped into short cube-like pieces. The dimensions of the granules are controlled by the thickness of the strip, the distance separating the slitter discs and the rotational speed of the cutter in relation to the rate of feed.

The stair-step dicer consists of a series of serrated knives mounted on a rotor which mesh with similarly serrated fixed knives protruding into a cutting chamber. The strip material is fed diagonally to the knives by means of a series of drive rolls and the material is cropped into uniform cubes. The size of the cubes in this case is controlled by the thickness of the strip, the pitch of the teeth on the cutting knives and the rate of feed in relation to the rotor speed (*Figure 10.8*).

Figure 10.8.
A small vertical-feed stair dicer
with the cover open to show
the serrated rotary knives.
(Courtesy U.S. Industries
Inc. Engng Co. Ltd)

Compared with the granules from a rotary cutter, diced material is considerably more uniform and has a higher bulk density. The material still tends to include a small proportion of fines, however, depending on the sharpness or otherwise of the knives, and also sometimes contains 'longs' due to incomplete severing of granules. The slitter-type cube cutters are said to be more prone to these defects than the more elaborate stair-step type.

The remaining methods of granulation from an extruder both use multi-orifice dies to produce a number of strands or filaments of material rather than a single strip. The methods of producing granules from these strands differ, however, and are known respectively as 'strand' and 'die-face' cutters.

10.5.3 STRAND CUTTERS

In this method the strands, which may be up to 30 or more in number, are taken from the die into a cooling bath after which they are dried by an air wiper. From the drying operation they are fed into the cutter by means of a nip roll assembly, where rotating knives crop them to the required length (*Figure 10.9*). The shape of the granules is dependent on the profile of the die orifices and their size is controlled by the degree of draw-down and the rotational speed of the knife in relation to feed rate.

Figure 10.9. Strand cutter or pelletiser with cover removed showing one knife on the rotor. (Courtesy Blackfriars Engng Ltd)

Strand cutters, when operating correctly, produce very few fines and no longs. Trouble is sometimes experienced, however, in keeping the strands separate and in ensuring that they are all of uniform size and are fed to the cutter at uniform rate.

10.5.4 DIE-FACE CUTTERS

Instead of feeding the strands through a water bath and nip rolls into a cutter located at a considerable distance from the extruder, rotating knives may be mounted at the die face to crop the strands into short lengths immediately they emerge from the die. In the majority of systems die-face cutters consist of one of more spring-loaded flying knives which wipe across the die face and with most thermoplastics elaborate systems of cooling are required to chill the granules immediately they are cut to prevent them sticking together (*Figure 10.10*).

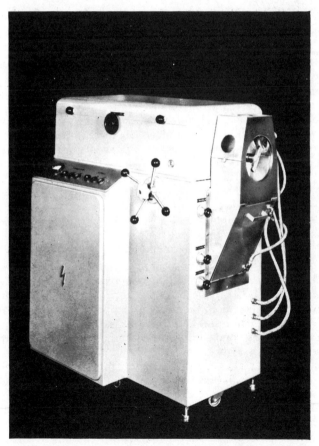

Figure 10.10. General view of a die face cutter (Courtesy Gerhard Kestermann)

From the die face the granules are taken by a cooling system on to a vibrating or other conveyor where they are separated from the cooling water and continuously dried by residual heat.

The size of die-face-cut granules is determined by the size of the die orifices and the rate of extrusion in relation to the speed of rotation of the flying knives, which are frequently equipped with variable speed drive.

Granules produced by die-face cutting techniques are free from both fines and longs and have the highest packing density. Except for small, low production operations, however, the equipment is often difficult to operate and is elaborate and costly.

10.6 Scrap reclamation

The reworking of scrap materials and reject products is an essential part of the economics of all processes for the manipulation of thermoplastics. In blow-moulding operations the heads, tails and necks and other trimmings must be re-used and in injection moulding the sprues and runners are frequently re-ground beside the machine for immediate re-feed. In sheet forming operations the percentage of scrap may be as high as 50 per cent or more.

In extrusion processes the waste material which occurs at start-up and before correct dimensions and finish are obtained must be re-used as must other reject material. Frequently products which are sold at very low price—such as some grades of garden hose—are made entirely, or almost entirely, from scrap material, whilst others contain a varying proportion of scrap depending on such factors as colour requirements and end use.

Scrap material frequently cannot be re-used in the process on which it was originally formed, but is reserved for use in less critical products. The scrap produced in the production of clear films, for example, can seldom be re-used for this purpose, but will be reserved for inclusion in garden hose or in the production of non-critical profiles, second-grade pipe or in the injection moulding of housewares such as dustbins, etc. The tremendous growth of the thermoplastics processing industry during the past decade has resulted in the formation of a number of specialist firms whose business it is to deal in scrap materials and to reprocess them for sale as second-grade materials at reduced price. The problems which arise in the reprocessing of a wide range of scrap thermoplastic as a specialist operation are very difficult and involve such matters as the segregation of colours, the complete separation of one thermoplastic from others with which it may be contaminated, to say nothing of scrap cable with cotton markers, fabric-backed sheet and laminates.

The best of the reprocessing firms have their own systems for dealing with these problems, which are obviously outside the scope of this book. On the other hand, as the reprocessing of scrap frequently involves the use of extruders, a few words on the subject will not be out of place.

The reject material which arises during the normal course of plastic moulding or extrusion will be in many forms. It can vary for example, from scrap polyethylene film through scrap pipe, mouldings and vacuum formings in a range of sizes, to the massive heavy lumps which result when a large extrusion die is dismantled for cleaning.

10.6.1 SCRAP OF AVERAGE THICKNESS

Mouldings of medium wall thickness, tubes, pipes and average extruded profiles will, if free from contamination, be simply granulated on a rotary cutter and re-used in this form. Alternatively if the colour needs to be modified, an appropriate quantity of a suitable colour concentrate or even a dry colour will be blended with the regranulated material in a tumble mixer or ribbon blender.

10.6.2 FILM SCRAP

Reject film products, on the other hand, are so thin that when reground they do not feed correctly to an extruder screw and normally require some form of densifying operation before they can be satisfactorily re-used. The developments of suitable simple techniques for the handling of scrap films—particulary polyethylene films—without the use of costly plant is a difficult problem which at the time of writing has not been entirely resolved. In general three basic processes are used.

10.6.2.1 *Reclamation by mill or internal mixer*

After the normal initial sorting into grades and examination for contamination, etc., the film is dumped in rolls into an internal mixer where it is melted; from the internal mixer the melt may be passed directly into a heated hopper extruder strainer equipped to produce a strip of suitable dimensions or it may be fed to a two-roll mill where it is sheeted in the normal way. The strip or sheet material, which is usually about 3 mm (⅛ in) thick, may then be granulated or diced.

Alternatively, the scrap film may be fed directly to the two-roll mills, thus avoiding the use of a costly internal mixer.

Although both of these methods require expensive plant they do have the advantage of producing a material of high bulk density. A further advantage being that blending with other grades and pigmenting can be carried out without difficulty in the same equipment and at the same time.

10.6.2.2 *Granulation/fluidising mixer*

In this process the film is granulated in the normal way and then fed to a high speed fluidising mixer. The action of the high speed mixer causes the flake-like film particles to heat up and curl, thus improving their bulk density.

Although this process does effect some improvement and does not require costly plant, the bulk density of the material so produced is usually still comparatively low.

10.6.2.3 *Pellet mill*

This method was developed in the U.S.A. by the Sprout-Waldron Co[25] and the basic principle consists of granulating the film in the normal way and then feeding it to a 'pellet mill' of special design. The pellet mill consists of a cylindrical chamber with a large number of small holes pierced through its walls. Inside the chamber are rolls and spreading tools which rotate on a common axis running along the cylinder length. The granulated film is fed into the cylinder where the combined action of the rotating spreaders and rolls partly plasticises the material and forces it outwardly through the holes in the cylinder wall. The spaghetti so produced can either be cut off to pellet length by knives rotating around the external periphery of the cylinder or alternatively the cylinder can itself revolve against stationary knives.

Although the pellets produced by this process are denser than those produced by the fluidising mixer technique, they are still less dense than those produced by milling or with an internal mixer.

10.6.3 MASSIVE SCRAP

Massive lumps from extruder dies and spewings from injection machines, etc., also require special treatment. Such masses of material may contain degraded sections which must first be cut off and discarded. They may also be contaminated with foreign matter due to handling whilst hot and may even contain parts of broken-up screen packs and other metallic contamination.

When the degraded or contaminated portions have been removed the lumps are often cut up on a band saw into pieces of a convenient size for grinding on a rotary cutter or they may be passed to a mill or an internal mixer with other scrap material.

10.6.4 GENERAL VIEWS ON SCRAP MATERIALS

Before leaving this question of scrap a few words on the handling of reject material generally may be of value.

To the ordinary processor of thermoplastics, scrap material is often an embarrassment and always a nuisance and is seldom treated with the respect it deserves. A proportion—sometimes a high proportion—of the profit on a particular job goes into the scrap which is produced in start up and in obtaining size. That scrap must be collected very carefully and re-used before the full profit of the operation can be realised. Yet so often the scrap is allowed to fall on to a dirty floor to be picked up later, perhaps after it has been walked on by the operators, bagged, placed in a dirty scrap store and forgotten until such time as a dealer comes round to make a very low offer for the accumulation of waste material.

This is not the way to run a profitable extrusion business. Scrap material should always be kept off the floor and covered and an effort should be made to regrind and re-use it as soon as possible after it has been produced. If a central scrap store is used it should be adequately staffed and equipped so that there is no delay in the regrinding and proper bagging and cataloguing of reject materials. Needless to say, the scrap stores should also be kept clean! Even more efficient is the use of cooled granulators which can be placed below the extruder die and will accept the hot extrudate and granulate it in its uncontaminated condition.

REFERENCES

1   ECKERT & ZIEGLER, Germ. Pat. 467753 (16 March 1927)
2   COOPER, G., 'Low Pressure Injection Moulding', *S.P.E. J1*, **15**, 966 (1959)
3   ANON., 'The Troester Extr-a-Formatic Automatic Injection Moulding Machine', *Plastics, Lond.*, **23**, 371 (1958)
4   TENAPLAS LTD, unpublished laboratory notes
5   CELLULOID CORPN OF U.S.A., Br. Pat. 422232 (8 Jan 1935)
6   ECKERT & ZIEGLER, Br. Pat. 511764 (23 August 1939)
7   F.I.M.S.A.I., Br. Pat. 676602 (31 December 1949)
8   BECK, H., Germ. Pat. DT 858310 (4 December 1952)
9   HARTLEY, A.W.E., Br. Pat. 633568 (19 December 1949)
10 BRITISH THOMSON HOUSTON LTD, Br. Pat. 697273 (16 September 1953)
11 R.H. WINDSOR LTD, Br. Pat. 791557 (22 June 1956)
12 BAIGENT, K.H., Br. Pat. 760480 (4 June 1954)
13 TUBE TURNS PLASTICS INC., Br. Pat. 808369 (2 July 1956)
14 FISHER, E.G. and MASLEN, W.A., 'Preplasticizing in Injection Moulding', *Br. Plast.*, **32**, 417, 468, 516 (1959)
15 FISHER, E.G. and MASLEN, W.A., 'Blow Moulding Techniques and Equipment', *Int. Plast. Engng*, 1, 2, 60, 128 (1961)
16 FISHER, E.G., *Blow Moulding of Plastics*, Iliffe, London (1971)
17 FARBWERKE HOECHST A.G., Br. Pat. 863417 (3 September 1957)
18 PIRELLI S.P.A., It. Pat. 481022 (19 May 1953)
19 INJECTION MOULDING CO. (MAINES, J.R.) U.S. Pat. 2632202 (19 October 1950)
20 PLASTICS, S.A., Br. Pat. 793445 (11 October 1954)
21 OWENS ILLINOIS GLASS CO. OF THE U.S., Br. Pat. 767649 (25 March 1955)
22 KALAHAR, M.J., 'Advances in Continuous Vacuum Forming Processes', *Plast. Technol.*, **4**, 335 (1959)
23 ANON., 'Screw Caps Formed—Fast', *Mod. Plast.*, **46**, 58 (1969)
24 HEDWIN CORPN, U.S.A., Br. Pat. 821173 (18 March 1957)
25 BAMBERGER, G., 'Polyethylene Film Scrap a Problem? Try This', *Mod. Plast.*, **37**, 114 (1960)

The Extrusion of Thermosets

11.1 History and general

Early experimental work on the extrusion of thermosetting materials was undertaken in Germany in 1930. The process then used was a simple adaptation of the conventional compression moulding system. A hand-operated vertical press was used, in which was mounted an elongated, cylindrical die, open top and bottom, through which material was forced by a closely fitting ram attached to the moving platen of the press. Phenolic moulding material in pellet form was fed to the die and an operator applied the thrust necessary to force the material through the die by turning a large handwheel. Many difficulties were encountered on this primitive extruder but a few feet of more or less uniform product were obtained. This method did not allow the continuous extrusion of unlimited lengths of material because of the interference of the floor with the downward movement of the product and this difficulty quickly terminated the use of a vertical press arrangement.

Late in 1930, Peter Kopp, also working in Germany, took out a British patent on a process in which articles were moulded continuously on a press arrangement similar to that described above[1]. Moulding material was fed to the press in pellet form when the ram was completely withdrawn from the die and between each change of material a metal disc was introduced to prevent the discrete moulding bonding together. The method of utilising the frictional resistance of the articles passing through the die to build up a back-pressure was emphasised in the claim and this idea is fundamental to the process of extruding thermosetting materials. Later in the same patent, mention was made of the possibility of producing sections of any desired length by omitting the metal partitions. The patent then went on to describe how the separate charges would bond together if a temperature gradient were maintained through the apparatus from the feed to the discharge end. The extrusion process was thus carried out in a continuous series of steps to produce a homogeneous section of infinite length.

Further British patents were granted to Kopp, covering a discharge opening perpendicular to the ram axis and certain elaborations of the idea of building up back-pressure by means of the friction between the cured section and the die. The fact that the die need not be the same shape as the ram was set out in these subsequent claims, as well as the incorporation of braking devices and cooling on the feed zone to maintain the necessary temperature gradient throughout the apparatus.

During the early 1930s thermosetting extrusion techniques developed quite rapidly in Germany, and horizontal reciprocating presses were built to operate automatically. Interest began to awaken in the U.K. and certain rights under the early German patents were acquired by British concerns, with a view to carrying out experimental work and subsequent production on a commercial scale. A prominent U.K. company in this early period of thermosetting extrusion development being Gestetner Ltd, who, in 1934, installed what is believed to be the first commercial machine, purchased from Werner Pfleiderer of Germany. The interesting potential which thermosetting extrusions appeared to have at that time aroused interest in the U.S.A. where development progressed along similar lines[2].

Despite its early promise, this particular field of extrusion, although a thriving business on a small scale in many countries, is today unimportant compared with its thermoplastics counterpart[3,4]. The development of new thermosetting materials, in particular glass-fibre-reinforced polyester resins, has, however, resulted in a revival of interest in the extrusion of thermosets, albeit, in a rather different form, which has led to the establishment of a further very important branch of the extrusion industry in the production of profiles, pipes and rods, etc., in these special thermosetting materials. This specialised process will be dealt with in a later section.

11.2   Basic principles of the process

Thermosetting resins, as their name implies, can be hardened and set by submitting them to a heating process. During the heating cycle they soften and can be moulded or extruded, but on further heating under pressure undergo an irreversible chemical change and once they have set they cannot be re-softened by subsequent heating. These materials are, therefore, completely different from thermoplastics which may be repeatedly re-softened by the application of heat.

Because of the non-reversible behaviour of thermosetting materials during a heating cycle the methods used to mould or extrude them differ somewhat from those which are suitable for thermoplastic materials. In order to obtain good quality products from thermosets it is necessary not only to heat the material but also to subject it to a pressure of several tonf/in² (1 tonf/in² $\simeq$ 15 MN/m²). Such high pressure requirements increase the difficulty of using screw extruders to deal with these materials. Furthermore, the possibility that the material might heat-harden owing to exotherm within the mechanism, with the resultant difficulties of dismantling and cleaning the screws, makes it difficult to use the continuous screw extrusion technique.

Largely because of the high pressures required, thermosetting materials are normally extruded on machines which are essentially reciprocating hydraulic presses. In the case of phenolic materials the resin in dry powder or pellet form is fed to the machine from a hopper arranged above a feed aperture in the water-cooled punch, or feed chamber, in much the same way as the feed of an injection moulding machine. The press ram causes the punch to close the feed aperture on its forward stroke and compacts the material against the preceding charges. This process is repeated at each stroke of the punch and material gradually progresses through the heated die, undergoing changes in profile and temperature at considerable pressures. The main change in die shape from the circular cross section of the punch to the form of the required section takes place in the flow zone. At this stage the material is completely softened and compressed to its final shape, after which it enters a substantially parallel curing zone. The material is still plastic at this stage but begins to cure as it enters the final section, which is very slightly tapered to accommodate the shrinkage which accompanies curing (*Figure 11.1*).

The pressure exerted by the punch is built up against the frictional resistance offered to the section in the curing zone of the die which is in the region of 225-300 mm (9-12 in) in length. If it is impossible to build up the necessary pressure in this way, additional control may be exercised by means of a die chuck, which is, in effect, a type of collet capable of increasing the control pressure of the die wall on the section.

When a hollow section is extruded, the mandrel which forms the centre profile must be supported in the correct position with respect to the outer profile and, furthermore, the supports must be capable of withstanding the tremendous end thrusts from the pressure in the material. The mandrels for small tube dies are supported on legs in much the same way as the mandrel of the tube die described in Chapter 8; these supporting fins are located in the flow zone of the extrusion machine. When sections with very large centre mandrels are being produced, other supporting methods are used since the force on a mandrel of large cross section would easily shear the small legs. One method consists of using the supporting fins merely for positioning, thereby reducing their cross section, and extending the mandrel back through a hollow punch to a rigid crosshead.

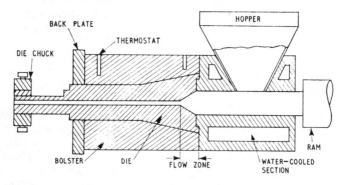

Figure 11.1. Diagram of extrusion process for thermosetting materials. (Courtesy British Resin Products Ltd)

The efficient operation of the machine obviously depends to a very great extent on the suitability or otherwise of the die. This component is made of high quality nickel chrome tool steel, highly polished and chromium-plated on its working faces; and with reasonable care and periodic replating such a tool will produce many hundreds of thousands of metres of section. These dies, moreover, as mentioned above, are much longer than those used in thermoplastics extrusion, with the result that they are much more difficult to manufacture. Simple symmetrical sections can often be bored from a solid billet but dies for more complex shapes are normally made in two halves. It is then necessary to ensure that the two parts of the die match correctly and that they are held together with no gap between their mating surfaces so that flash is avoided. The die itself is fitted inside a sleeve or chase, which is heated externally by electrical band heaters controlled by thermocouples in zones. This zonal control enables the required temperature gradient to be maintained through the plastics material from the water-cooled feed to the die orifice.

As in the case of injection-moulding machines, there is a certain minimum pressure which must be exerted on the material by the ram and a certain relationship must, therefore, be maintained between the cross-sectional area of the ram and that of the extruded section. This ratio is called the 'punch ratio', and its magnitude has a considerable bearing on the quality of the resulting product.

Figure 11.2. A modern thermosetting extrusion press shown in operation on the production of phenolic tube. (Courtesy Gestetner Ltd)

An extruder for thermosetting materials as referred to above is essentially a horizontal press and consists of two columns with a fixed crosshead which carries the dies and a moving table carried on the columns to which the die punches are attached. The main rams and the withdrawing rams are connected through automatic control valves to a hydraulic system. Most modern presses operate automatically and incorporate facilities for controlling the pressure and varying the length of the punch stroke. Such a press is shown in operation in *Figure 11.2*.

11.3 Differences and difficulties compared with thermoplastics extrusion

The differences in the behaviour of thermoplastics and normal thermosetting materials during processing and their different subsequent properties were outlined at the beginning of Section 11.2. It is these differences which prevent the two types of material being worked in the same way and, in particular, it is the non-reversible change through which thermosetting materials pass which limits the methods of producing sections in this material. The range of products which may be extruded in thermoplastics materials, from film to solid sections, is vastly superior to the range of sections possible with thermosets and it does not seem likely that this situation will change. In fact, judging by the rate at which new thermoplastics materials are being introduced, the disparity is likely to increase.

By virtue of the fact that little change can be made to the shape of a thermoset product when once it has left the machine a separate die is necessary for each individual section and, moreover, the die must accurately correspond to the required profile. On the other hand, great changes in section may be effected in the extrusion of thermoplastics after the material has left the die. This lends great versatility and flexibility to this process and, as a result, the dies for thermoplastics need not be accurately matched to the required section since there are many possibilities of changing and correcting the profile after it leaves the die. However, once a die has been made for a thermosetting section the product is independent of screw pulsing, take-off variations, cooling problems, and the many other difficulties which beset the thermoplastics worker. Furthermore, it cannot be denied that, in general, the dimensional consistency obtainable from a thermoset extruder is greater than that from a thermoplastics machine.

There is a great difference, however, in the cost of dies to make comparable sections by the two processes. The length of die which has to be accurately machined is much greater for thermosetting applications and because of the higher pressures involved tougher materials must be used with attendant machining difficulties.

The rates at which sections can be produced by the two processes differ widely. A thermoplastics extruder can produce a reasonably consistent product at rates which can be measured in centimetres per second and in the last resort this may possibly be determined by the rate at which the product can be cooled. Sections made from thermosetting materials, on the other hand, are produced in millimetres per second and the speed of this process is limited by the rate of cure, which of course cannot be widely varied. Special grades of thermosetting materials have been developed with the object of speeding up the cure time and rate of production.

*Figure 11.3.
A selection of stock rigid
extrusions from Rockite
phenolic materials.
(Courtesy British
Resin Products Ltd)*

Despite the disadvantages and difficulties of thermoset extrusion which have been outlined above, it seems likely that there will always be a demand—although small—for extruded sections in thermosetting plastics, as these materials have characteristics which are not so far possessed by the common thermoplastics. In addition to their excellent electrical properties and reasonable chemical resistance, some thermoset extrusions will withstand temperatures of 110 °C continuously, and up to, 160 °C for short periods, and their structural rigidity is not impaired by working at these high temperatures. Furthermore, the extrusions as produced are dimensionally stable, have good finish, and can be made to very close limits ready for use directly from the extrusion machine.

A selection of stock profiles extruded from thermosetting phenolic compounds is shown in *Figure 11.3*.

11.4   Extrusion of reinforced plastics

Although many of the thermosetting resins, when correctly compounded and processed in the moulded form, have good dimensional stability and stiffness, excellent heat resistance and other properties, they are usually limited in impact strength. It is common practice, therefore, to include some form of fibrous reinforcing material in thermosetting mouldings which are intended for high impact applications. The fibrous reinforcing materials in widespread use for this purpose include chopped rag, asbestos and cellulose fibres, paper, cotton, ramie, sisal, etc.

One of the most important new materials available for the reinforcement of resins is glass fibre, either as continuous filament, chopped fibre, woven cloth, or in other forms. This reinforcement offers several obvious advantages over the organic or mineral fibres referred to above. Glass fibres have exceptionally high tensile strength and Young's modulus, are fireproof and immune to microbiological and most forms of chemical attack, and have good weathering properties in addition to high dimensional stability and heat resistance.

Unfortunately, however, glass fibres have poor resistance to transversely applied abrasive affects, and the early attempts to use this material as a resin reinforcement resulted in failure due to the crushing or abrasion of the fibres during high pressure moulding. The development of polyester and other thermosetting resins which could be moulded at low pressure completely overcame this difficulty and allowed the outstanding strength and other properties of glass fibres to be fully utilised. The manufacture of glass-fibre-reinforced plastics in various forms and by a variety of processes, including a form of continuous extrusion, is now an important branch of the plastics industry.

Polyester resins as used for low pressure moulding are organic compounds containing unsaturated groupings which can be polymerised either alone or with other unsaturated monomers to give cross-linked infusable resins. The resins as commercially available, are usually in the form of syrups and the unsaturated cross-linking monomer is often styrene, although there is a formidable list of possible alternatives. The polymerisation reaction is initiated by means of heat combined with an organic peroxide catalyst or at room temperature by means of additional accelerating agents such as cobalt naphthenate or dimethylaniline. The manufacture of glass-fibre-reinforced polyester structures, therefore, consists basically of mixing the polyester resin syrup with the correct proportion of cross-linking monomer and then adding the·

catalyst to the mixture and finally the accelerator if used. This mixture is then applied to the glass fibre and allowed to set into rigid structures. This series of steps thus gives a clue to the processing methods available for the so-called continuous extrusion of glass-fibre-reinforced polyester resin materials.

11.4.1 PROCESS USING CONTINUOUS GLASS FILAMENTS OR ROVINGS

The original polyester/glass continuous extrusion process—now known as 'pultrusion'—consisted of drawing a bunch of continuous filaments through a bath of catalysed resin-monomer mixture and then through a die to set the shape of the extrusion and to control the resin content by squeezing out any excess. From the forming die the shaped extrusion passed through an oven where final cure was effected.

It soon became apparent that the commercial exploitation of the above process was dependent upon the rate of cure of the resin, which determined the rate of production, and in order to achieve an economic output of finished product extremely long curing ovens were required. This led to the development of a high frequency curing system by Glastrusions Inc. and Goldsworthy Engineering Inc. in the U.S.A., which utilised a PTFE die and a high frequency field to replace the lengthy radiant heat ovens[5-7]. This process has been successfully used for the manufacture of glass-reinforced products such as sheet, pipe and guttering for the building trade, structural beams, etc., in addition to solid rod stock for various applications[8].

11.4.2 PROCESS BASED ON RANDOM DISTRIBUTION OF ROVINGS

The extrusion of continuous-filament reinforced polyester resins has certain disadvantages, principally that the quality of the extrusion is dependent on the viscosity of the resin-catalyst mixture which determines the effectiveness of the impregnation or wetting and also controls the glass/resin ratio, both of which affect the properties of the finished product. Articles produced by this process are also limited in that their reinforcement is in one direction only, i.e. parallel to the length. This has led to the development of an improved process for the continuous extrusion of reinforced plastics into tubes, rods, pipes and various other profiles in which it is claimed that the glass reinforcement can be tailored to obtain optimum strength in any required direction or directions[9,10]. In this method the fibrous reinforcement may be in the form of woven strips, random mat or other form and is drawn through an impregnating bath of prepared resin and thence through a heated shaping and confining die[11]. Some of the profiles which have been produced by this method are shown in *Figure 11.4.*

Figure 11.4. A selection of profiles produced by a recently developed extrusion process for reinforced plastics. (Courtesy English Electric Co. Ltd., Reinforced Plastics Division)

11.4.3   DOUGH MOULDING COMPOUNDS

With the introduction of polyester dough moulding compounds—known as 'bulk-moulding compounds' or 'BMC' in the U.S.A.—or glass/resin premixes as they are also sometimes called, further possibilities have arisen for the extrusion of reinforced plastics.

Dough moulding compounds are usually of a putty-like consistency consisting of resin, catalysts, fillers, colouring materials combined with glass, sisal or other fibrous reinforcements in relatively short lengths 6-50 mm, (¼-2 in). These compounds have been successfully extruded in the uncured form as raw material stock for moulding, on both piston and screw machines, by British Industrial Plastics Ltd[12], and into a range of finished profiles.

11.5 Conclusion

Although the quantity of reinforced plastics being produced by extrusion systems at the present time is relatively small compared with other reinforced plastics processes, or particularly with thermoplastics extrusion, it is apparent that the process has considerable potential. The commercial development and exploitation of reinforced plastics processes generally has been retarded and difficult because of the lack of a straightforward and simple mass production system. A satisfactory continuous extrusion process which is now becoming available may well provide the future answer to this difficulty.

REFERENCES

1 KOPP, P., Br. Pat. 380824 (26 September 1932)
2 SIMONDS, H.R., WEITH, A.J. and SCHACK, W., *Extrusion of Plastics, Rubber and Metals*, Reinhold, New York (1952)
3 DAVIES, D.N., *Plast. Prog.*, 235 (1951)
4 OATES, A.E., 'Developments in Thermosetting Extrusions', *Rubb. Plast. Age*, **35**, 35 (1954)
5 ANON, 'New Machine Produces Reinforced Plastics', *Aust. Plast.*, **10**, 23 (1954)
6 HENNINGS, A.R. and CUMING, G.B., in *Glass Reinforced Plastics*, 2nd edn, Ed. Morgan, P., Iliffe, London (1952), p. 152
7 HAGEN, H., 'The Manufacture of Polyester Pipes, Fittings, Rods and Sections', *Kunststoff-Rdsch.*, **3**, 1 (1956)
8 ANON., 'For Pultrusion 1974 Looks Like the Best Year Yet', *Mod. Plast. Int.* **4**, 46 (1974)
9 ANON., 'Continuous RP Sections Produced Automatically', *Mod. Plast.*, **40**, 93 (1962)
10 MORRIS, D.R., 'A Sausage Machine for Reinforced Plastics', *New Scient.*, **16**, 509 (1962)
11 UNIVERSAL MOLDED FIBER GLASS CORPN, Br. Pat. 908753 (16 January 1961)
12 NEWTON, J., 'New Developments in Polyester Dough Moulding Compounds', *Plastics, Lond.*, **26**, 51 (1961)

Appendix

List of symbols used in this book; the definitions are standard throughout this work, except where otherwise stated in the context:

A = a collection of three terms from the full force balance equation for solids conveying

a = ratio of pressure flow to drag flow

b = axial channel width

c = length of q, see below

D = screw diameter

d = density

E = screw eccentricity factor

e = axial flight width

F_D = shape factor for drag flow

F_P = shape factor for pressure flow

G = efficiency of motor

h = channel depth

K = ratio of forces acting perpendicular to reference plane to those acting parallel to it

k = constant

L = screw length

M_L = volume of flight clearance

M_S = volume of discrete section in twin-screw extruder

N = screw speed

n = number of parallel screw channels

p = pressure

Q = flow rate—output

Q_F = output for feed section

Q_D = drag flow rate or volume

Q_L = leak flow rate or volume

Q_p = pressure flow rate or volume

q = fraction of screw thread pitch line unobscured by meshing in twin-screw extruder

R = residence time

t = screw-lead = $D \tan \phi$

U = velocity of barrel relative to screw = $\pi DN \cos \phi$

u = velocity of a specific particle (*see* context)

V = peripheral speed of root diameter of screw

v = velocity

v_D = motion due to drag flow

v_p = motion due to pressure flow

w = channel width = $b \cos \phi$

Z = power requirements

Z_p = power requirement for pressure

z = distance along helical axis

$\dfrac{\partial p}{\partial l}$ = axial pressure gradient along screw

Δ_p = pressure drop along screw

Δ_{pD} = pressure drop through die

Δ_Q = output differential due to surging

α = constant for drag flow ·

β = constant for pressure flow

δ = flight land clearance

η = viscosity

η_L = viscosity in land clearance

θ = angle of movement of outer surface of solids plug

μ = coefficient of friction

ρ = angle of movement due to friction of material against screw body

τ = shear stress

ϕ = helix angle

ϕ_1 = helix angle at core diameter

ψ = dissipation function for viscous shear energy

Index